Recycled Pulsars

Bryan Anthony Jacoby

DISSERTATION.COM

Boca Raton

Recycled Pulsars

Dissertation.com
Boca Raton, Florida
USA • 2008

ISBN-10: 1-58112- 393-0
ISBN-13: 978-1-58112-393-7

Recycled Pulsars

Thesis by

Bryan Anthony Jacoby

In Partial Fulfillment of the Requirements

for the Degree of

Doctor of Philosophy

California Institute of Technology

Pasadena, California

2005

(Defended December 8, 2004)

Acknowledgements

I would like to thank some of the many people who have influenced and contributed to me and this work. Those who have my gratitude are too many to thank here, but I will do my best. But first, I would like to acknowledge the generous financial support of the NSF Graduate Research Fellowship Program and the NRAO GBT Student Support Program that made this work possible.

My mother has always encouraged my curiosity and supported my interest in science, even when it involved sacrifice on her part. I hope that I will be as good a parent.

Alex Wolszczan, along with Stuart Anderson and Brian Cadwell, introduced me to the fascinating world of pulsars and radio instrumentation as an undergraduate at Penn State. I am lucky to have had Stuart as a willing mentor for the past decade.

I am grateful to my advisor, Shri Kularni, for providing the direction and resources for this research, and for his insights which continue to amaze me. Matthew Bailes has served as unofficial co-advisor and treated me like a member of his group from the beginning; I have learned much from him.

I have had the privilege of learning from several other exceptional astronomers. My time with Don Backer has been a true highlight, both personally and educationally. Marten van Kerkwijk, Deepto Chakrabarty, and Dale Frail have been generous and patient as well.

The members of the Swinburne pulsar group (past and present), namely Steve Ord, Aidan Hotan, Haydon Knight, Willem van Straten, Russell Edwards, and Craig West, have become not only valued colleagues but also very good friends. Along with the many other people I have had the good fortune of knowing at Swinburne, Parkes, and Green Bank, they have made my long stints away from home much more pleasant as well as productive. I have spent less time observing at Palomar but it is one of my favorite places, largely because of the exceptional people who work there.

So many of my fellow grad students at Caltech have helped me in some way over the past several years that it is hopeless to try to thank them all. I have shared an interest in neutron stars with David Kaplan, Adam Chandler, and Rick Jenet, and fear that I have learned more from them than I managed to teach in return. Brian Cameron, a recent addition to the team, has been a pleasure to work with; I greatly appreciated his willingness to help with observing and other responsibilities in the stressful late stages of writing this thesis. Edo Berger has kindly helped me several times with data analysis. I also thank Jon Sievers, Pat Udomprasert, Cathy Slesnick, Melissa Enoch, Josh Eisner, George Becker, and Stuartt Corder for being accommodating and enjoyable office mates. My classmates Mike Santos and Matt Hunt are great friends and are two of the most interesting people I know. Mike, along with Rob Simcoe and Micol Christopher, helped to provide an outlet for me through their hard work captaining the Cataclysmic Variables, our department soccer team. Thanks to people here in Robinson would be incomplete without mentioning our crack team of systems administrators, Patrick Shopbell, Cheryl Southard, and Anu Mahabal, without whom very little would get done around here. John Yamasaki's work on pulsar instrumentation was critical to my research and I have missed him since he left Caltech.

This thesis is dedicated to my wonderful wife Megan, whose love and support have sustained me throughout this work.

Abstract

We present the results of a large-area survey for millisecond pulsars (MSPs) at moderately high galactic latitudes with the 64-m Parkes radio telescope, along with follow-up timing and optical studies of the newly-discovered pulsars and several others. Major results include the first precise measurement of the mass of a fully recycled pulsar and measurement of orbital period decay in a double neutron star binary system allowing a test of general relativity along with improved measurements of the neutron star masses.

In a survey of $\sim 4,150$ square degrees, we discovered 26 previously unknown pulsars, including 7 "recycled" millisecond or binary pulsars. Several of these recycled pulsars are particularly interesting: PSR J1528−3146 is in a circular orbit with a companion of at least $0.94\,M_{\odot}$; it is a member of the recently recognized class of intermediate mass binary pulsar (IMBP) systems with massive white dwarf companions. We have detected optical counterparts for this and one other IMBP system; taken together with optical detections and non-detections of several similar systems, our results indicate that the characteristic age $\tau_c = P/2\,\dot{P}$ consistently overestimates the time since the end of mass accretion in these recycled systems. This result implies that the pulsar spin period at the end of the accretion phase is not dramatically shorter than the observed period as is generally assumed. PSR J1600−3053 is among the best high-precision timing pulsars known and should be very useful as part of an ensemble of pulsars used to detect very low frequency gravitational waves. PSR J1738+0333 has an optical counterpart which, although not yet well-studied, has already allowed a preliminary measurement of the system's mass ratio. The most significant discovery of this survey is PSR J1909−3744, a 2.95 ms pulsar in an extremely circular 1.5 d orbit with a low-mass white dwarf companion. Though this system is a fairly typical low-mass binary pulsar (LMBP) system, it has several exceptional qualities: an extremely narrow pulse profile and stable rotation have enabled the most precise long-term timing ever reported, and a nearly edge-on orbit gives rise to a strong Shapiro delay signature in the pulse timing data which has allowed the most precise measurement of the mass of a millisecond pulsar: $m_p = (1.438 \pm 0.024)\,M_\odot$. Our accurate parallax distance measurement, $d_\pi = (1.14^{+0.08}_{-0.07})\,\mathrm{kpc}$, combined with the mass of the optically-detected companion, $m_c = (0.2038 \pm 0.0022)\,M_\odot$, will provide an important calibration for white dwarf models relevant to other LMBP companions.

We have measured the decay of the binary period of the double neutron star system PSR B2127+11C in the globular cluster M15. This has allowed an improved measurement of the mass of the pulsar, $m_p = (1.3584 \pm 0.0097)\,M_\odot$, and companion, $m_c = (1.3544 \pm 0.0097)\,M_\odot$, as well as a test of general relativity at the 3% level. We find that the proper motions of this pulsar as well as PSR B2127+11A and PSR B2127+11B are consistent with each other and with one published measurement of the cluster proper motion.

We have discovered three binary millisecond pulsars in the globular cluster M62 using the 100-m Green Bank Telescope (GBT). These pulsars are the first objects discovered with the GBT.

We briefly describe a wide-bandwidth coherent dedispersion backend used for some of the high precision pulsar timing observations presented here.

Contents

List of Figures

List of Tables

Chapter 1

Introduction

The story of the study of pulsars begins in 1054 A.D., when Chinese astronomers observed the appearance of a bright object in the sky, the Crab supernova. We now know that supernovae like the Crab are the death cries of massive stars. Nearly 900 years later, Baade & Zwicky (1934) hypothesized that a very compact, dense remnant — a dead star composed almost entirely of neutrons — could be produced in a supernova explosion. However, it was assumed that these small neutron stars would not emit enough thermal radiation to be detected by optical telescopes. In 1967, Jocelyn Bell was observing interplanetary scintillation of celestial radio sources and found periodic bursts of radio noise unlike anything that had ever been seen before. These pulses, repeating every 1.3 s and visible each day in the same part of the sky, turned out to be the pulsar PSR B1919+21 (Hewish et al., 1968), a spinning, highly-magnetized neutron star sending out beams of radio waves along its magnetic axis. When these beams sweep past the earth, we see the pulsar blink on and then off, much like a rotating lighthouse beacon. Soon thereafter, Staelin & Reifenstein (1968) discovered a pulsar spinning 30 times per second at the heart of the Crab supernova remnant.

The discovery of the first binary pulsar, PSR B1913+16 (Hulse & Taylor, 1975), and of the first millisecond pulsar (MSP), PSR B1937+21 (Backer et al., 1982), ushered in the era of recycled pulsars and high-precision timing. After a neutron star is born, it gradually slows and eventually ceases to shine. In some circumstances, the neutron star can be recycled and begin a new life by accreting matter and angular momentum from a binary companion, spinning the neutron star up to fantastic rates. To the chagrin of many pulsar astronomers, after over twenty years of searching for more MSPs, PSR B1937+21 is still the most rapidly rotating pulsar known, spinning over 640 times per second! MSPs, with more than the mass of the sun packed into an object the size of a small city, a magnetic field 10^8 times stronger than that of the earth, and spinning faster than the blades of a kitchen blender or the crankshaft of a Formula 1 race car engine, are among the most exotic objects known in the universe. Well over 100 recycled pulsars are known today (out of over 1,500 pulsars total) — many in globular star clusters.

With incredible rotational stability rivaling that of atomic clocks, recycled pulsars provide ideal laboratories for a variety of physical experiments. Binary pulsars with massive companions can be used to test theories of gravity (Taylor & Weisberg, 1989; Stairs et al., 2002), and by timing the most stable MSPs, it will be possible to constrain or possibly even detect low-frequency gravitational waves emitted by massive black hole binary systems (Jaffe & Backer, 2003; Lomen et al., 2003). The ability to conduct these ultra-precise experiments with stars scattered across the galaxy with precision far exceeding almost every other type of astronomical observation comes from the technique of pulsar timing. Briefly, the arrival times of pulses obtained from a given set of observations are compared with a model or mathematical description of the expected pulsar behavior, and in turn, used to improve our knowledge of the model parameters. The power of this technique comes from the ability to count exactly how many rotations of the pulsar have taken place between one observation and the next over a long period of time, leading to very small fractional uncertainties in the model parameters (see Bell, 1998 for a detailed overview). A large part of the motivation for present and future pulsar search efforts is finding objects that will improve the quality of these timing experiments, or allow completely new experiments as in the case of the recently-discovered double pulsar system (Burgay et al., 2003; Lyne et al., 2004). Another primary motivation is the detection of an object spinning substantially faster than PSR B1937+21, whose very existence would provide direct constraints to theoretical models describing the mass – radius relation of nuclear matter (Fig. 2.11).

Motivated by these considerations, we have surveyed a large area of the sky at moderately high galactic latitudes for pulsars (Chapter 2). This effort resulted in the discovery of 7 recycled pulsars (Chapters 3 and 4), including one remarkable object which has allowed the first precise measurement of the mass of an MSP (Chapter 5). We have completed a series of optical observations of pulsars with massive white dwarf binary companions (Chapter 6), one of which was discovered in the pulsar survey. Through long-term timing observations of the double neutron star binary PSR B2127+11C in the globular cluster M15, we have verified that general relativity provides a correct description of gravity at the 3% level, improved upon the previous mass measurements of the pulsar and its companion, and verified the proper motion of M15 (Chapter 7). We have discovered three binary MSPs in the globular cluster M62 (Appendix A) as part of a study of globular cluster pulsars which is still in progress. Finally, we describe the design of an observing system used for some of the observations presented here (Appendix B).

Chapter 2

A Large-Area Survey for Radio Pulsars at High Galactic Latitudes†

B. A. JACOBY[a], M. BAILES[b], S. ORD[b], H. KNIGHT[b,c], A. HOTAN[b,c], R. T. EDWARDS[c], AND S. R. KULKARNI[a]

[a]Department of Astronomy, California Institute of Technology, MS 105-24, Pasadena, CA 91125; baj@astro.caltech.edu, srk@astro.caltech.edu.

[b]Centre for Astrophysics and Supercomputing, Swinburne University of Technology, P.O. Box 218, Hawthorn, VIC 31122, Australia; mbailes@swin.edu.au, sord@swin.edu.au, ahotan@swin.edu.au.

[c]Australia Telescope National Facility, CSIRO, P.O. Box 76, Epping, NSW 1710, Australia; Russell.Edwards@csiro.au.

Abstract

We have completed a survey for pulsars at high galactic latitudes with the 64-m Parkes radio telescope. Observing with the 13-beam multibeam receiver at a frequency of 1374 MHz, we covered $\sim 4,150$ square degrees in the region $-100^{\circ} \leq l \leq 50^{\circ}$, $15^{\circ} \leq |b| \leq 30^{\circ}$ with 7,232 pointings of 265 s each. The signal from each beam was processed by a $96-\text{channel} \times 3\,\text{MHz} \times 2-\text{polarization}$ filterbank, with the detected power in the two polarizations of each frequency channel summed and digitized with 1-bit resolution every $125\,\mu$s, giving good sensitivity to millisecond pulsars with low or moderate dispersion measure. The resulting 2.4 TB data set was processed using standard pulsar search techniques with the workstation cluster at the Swinburne Centre for Astrophysics and Supercomputing. This survey resulted

†Part of a manuscript in preparation for publication in *The Astrophysical Journal*

in the discovery of 26 new pulsars including 7 binary or millisecond "recycled" pulsars, and re-detected 36 previously known pulsars. We describe the survey methodology and results, and present timing solutions for the 19 newly discovered slow pulsars.

2.1 Introduction

Since the discovery of the first radio pulsar (Hewish et al., 1968), a great deal of effort in radio astronomy has been expended searching for pulsars, with about 1,500 pulsars known today. Until the late 1990s, most pulsars surveys focused on the region of sky near the plane of our galaxy since that is where the density of pulsars is highest. Also, early surveys usually were conducted at frequencies around 400 MHz because of pulsars' steep radio spectra, and because the larger beam produced by a given telescope at lower frequencies allowed for more rapid coverage of the sky, or conversely, longer integration on a given point in the sky, given a region of sky to cover and a fixed allocation of observing time.

However, there are clear advantages to searching for pulsars at higher frequencies. Most notably, dispersion and scattering are mitigated, allowing for better time resolution and sensitivity to pulsars with large dispersion measures (DMs). Also, the galactic synchrotron background has a steep spectrum and contributes little to the total system temperature at higher frequencies. The final disadvantage of high-frequency surveys, namely the slow sky coverage resulting from a small telescope beam, has been overcome by the innovative 13-beam multibeam receiver package at the Parkes radio telescope (Staveley-Smith et al., 1996) and 13 accompanying analog filterbanks which provide a combined beam area on the sky about 25% larger than the 70-cm system used in the Parkes Southern Pulsar Survey (Manchester et al., 1996). The Parkes Multibeam Pulsar Survey covering $\sim 1,500$ square degrees within 5° of the galactic plane has used this system to great effect, roughly doubling the number of pulsars known before this survey began (Manchester et al., 2001; Hobbs et al., 2004).

While pulsars descend from short-lived massive stars which are born and die in the galactic disk, older pulsars have had time to migrate to moderate to high galactic latitudes. Millisecond pulsars (MSPs) which spin roughly a hundred times or more each second are in this older group, having lived a life as a "normal" pulsar, then later being "recycled" to very fast rotation rates by accreting matter from an evolved binary companion. It is these objects which promise to reveal the neutron star equation of state: discovering ever faster spinning pulsars constrains the size of neutron stars, and measuring effects predicted by Einstein's general relativity allows us to determine their masses. Although the density of pulsars is less at high galactic latitudes than in the plane of the galaxy, a shallow survey of this relatively neglected part of the sky provides an efficient means for discovering recycled pulsars, as demonstrated by the Swinburne Intermediate Latitude Pulsar Survey in a region between 5° and 15° from the galactic plane (Edwards & Bailes, 2001b; Edwards et al., 2001). Here, we describe the results of a 21-cm multibeam survey for pulsars at higher galactic latitudes between 15° and 30° from the galactic plane.

2.2 Observations and Analysis

Between 2001 January and 2002 December, we observed roughly 4,150 square degrees in the region $-100^{\circ} \leq l \leq 50^{\circ}$, $15^{\circ} \leq |b| \leq 30^{\circ}$ in 7,232 individual pointings of 265 s with the 64-m Parkes radio telescope at a frequency of 1374 MHz. Using the 13-beam multibeam receiver system, the signal from each beam was processed by a 96 MHz $\times$ 2 filterbank, the powers from each polarization pair summed, and the result 1-bit sampled at 125 μs intervals. The resulting 94,016 survey beams comprising 2.4 TB in total were written to 98 DLT tapes for later analysis. The observing and data processing procedures were virtually identical to those described by Edwards et al. (2001). Our observing hardware and methodology were identical to the Parkes Multibeam Pulsar Survey, except that our integration time per pointing was shorter (265 s compared to 2,100 s) and our sampling interval was half as long (125 μs compared to 250 μs).

The minimum detectable flux $S_{\rm min}$ for a radio periodicity search can be calculated by a modified form of the radiometer equation,

$$S_{\rm min} = \frac{\alpha\, \beta\, (T_{\rm rec} + T_{\rm sky})}{G\, (2\, \Delta\nu\, t_{\rm int})^{1/2}} \left(\frac{\delta}{1-\delta}\right)^{1/2}, \qquad (2.1)$$

where α is the threshold signal-to-noise ratio (S/N), $\beta \approx 1.5$ is an efficiency factor taking account of losses such as quantization error, G is the telescope gain, $\Delta\nu$ is the observed bandwidth, $t_{\rm int}$ is the integration time, $T_{\rm rec}$ and $T_{\rm sky}$ are the receiver and sky contributions to the system noise temperature, and δ is the effective fractional duty cycle of the periodic signal, including contributions from the sample period, dispersion smearing, finite DM search step size, and scattering. Figure 2.1 shows representative sensitivity curves for this survey. In our calculations, we have assumed that the pulsar signal contains $(2\,\delta)^{-1}$ equally significant harmonics in the frequency domain.

Our large data set was analyzed with the cluster of 64 Compaq Alpha workstations at Swinburne University of Technology's Centre for Astrophysics and Supercomputing. The filterbank data from each survey beam were padded with 32 empty channels with appropriate frequency spacing so that dispersion was a linear function of channel number in this 128-channel space. First, each channel was searched for strong, narrow-band radio frequency interference (RFI), and affected channels were masked. We then used the "tree" dedispersion algorithm (Taylor, 1974) to break the 128 channels into 16 sub-bands, each dedispersed at a range of different DMs. This sub-band data could then be efficiently dedispersed and summed to form one of 374 trial DMs up to a maximum $562.5\,\mathrm{pc\,cm^{-3}}$. Once the sub-band DM reached twice the diagonal DM of $17\,\mathrm{pc\,cm^{-3}}$ (when roughly one sample period of smearing occurs within an individual channel), adjacent samples were summed (thereby doubling the diagonal DM) and the process repeated until the maximum search DM was reached.

The periodicity search used an adaptation of the search software from the Parkes Southern Pulsar Survey (Manchester et al., 1996) and followed standard pulsar search procedures. Briefly, the Fourier transform was computed for a given trial DM and searched for harmonically related patterns with a fundamental frequency greater than $1/12\,\mathrm{Hz}$, summing up to 16 harmonics of a given frequency. The large list of possible candidates in the three-dimensional space of repetition frequency, number of harmonics summed, and DM was consolidated into a list of the best 99 suspects from each survey beam. Each of these was then subjected to a time-domain optimization in spin period (P) and DM. For each beam, the 48 most promising suspects from this optimization were saved for further scrutiny.

Candidate pulsars were selected from 1.2 million search suspects using a variety of fixed and adaptive filters, and finally, by human examination. Typically, the roughly 45,000 suspects from a given tape were considered together. Filters were constructed to eliminate known interference frequencies and to enforce criteria to eliminate other low-quality suspects such as short-period suspects with large relative DM error. The S/N threshold was set at 9 for suspects with $P \geq 20\,\mathrm{ms}$ and 9.5 for shorter-period suspects to help eliminate the many

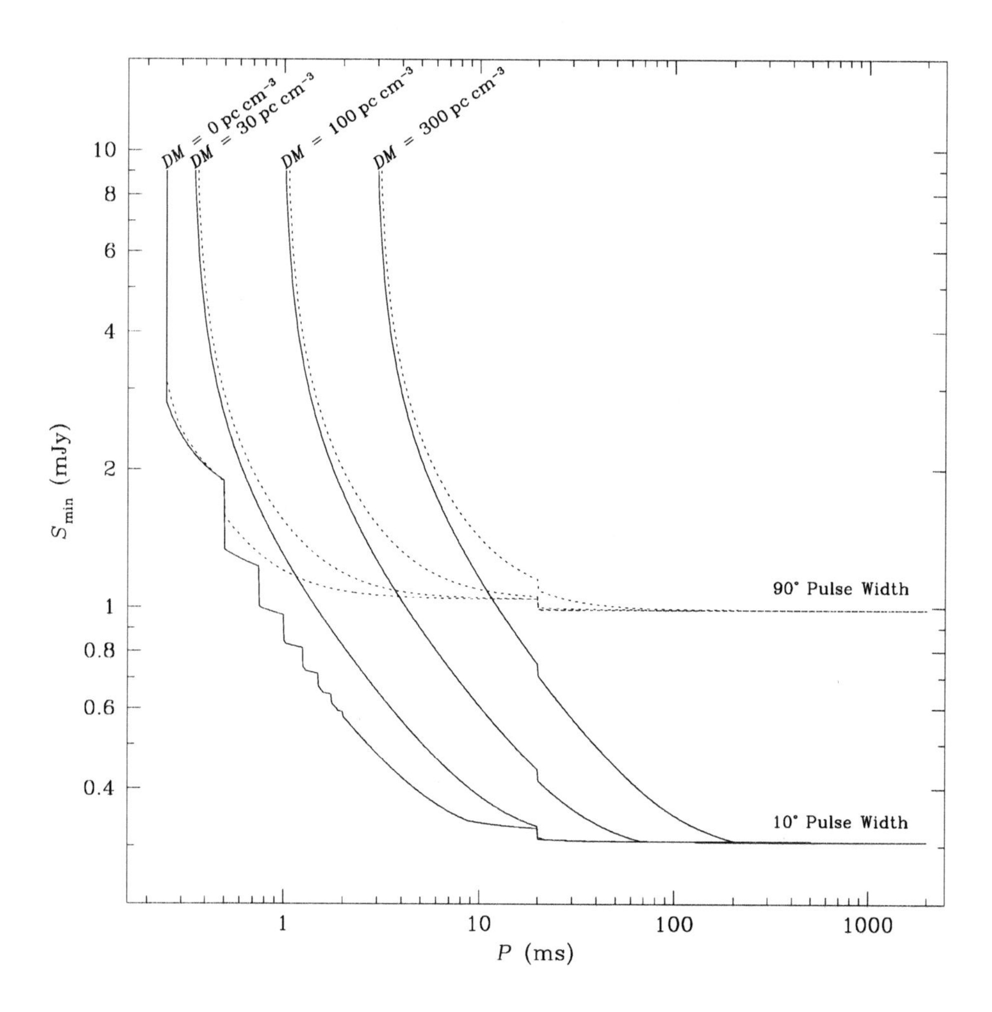

Figure 2.1: Survey sensitivity. Curves show minimum detectable 1400 MHz flux density as a function of spin period for pulsars with large (dotted curves) and small (solid curves) duty cycles for a range of dispersion measures as labeled.

spurious short-period signals found in our data. Finally, adaptive filters were used to allow signals that appeared at nearly the same period separated by many telescope beam widths many times on one tape. These automated parameter-based filters and screens allowed us to cull the suspect list to fewer than 10,000, each of which was then given much more human attention than would have been possible for the full set of suspects. Diagnostic information for two example candidate pulsars which were eventually confirmed is shown in Figures 2.2 and 2.3. Plausible pulsar candidates were re-observed, with the observation time depending on the strength of the candidate.

We note that in the case of one extremely convincing MSP candidate — later confirmed as PSR J1741+13 — it took four re-observations before it appeared again. This experience highlights the important role played by scintillation in pulsar surveys. Many other candidates which were not as convincing or interesting as this one were, of course, not afforded four confirmation attempts, and there are doubtless other pulsars in the survey region which were not visible at the time of the survey observation. We have no doubt that re-observing regions which have previously been surveyed will yield new discoveries, even when a new observing capability or strategy is not brought to bear.

2.3 Detected Pulsars

In addition to the seven recycled pulsars described in Chapters 3 and 4, this survey discovered 19 new slow pulsars. After confirmation, we began a roughly monthly timing program to determine phase-connected timing solutions for all slow pulsars using the Parkes 512×0.5 MHz $\times 2$ filterbank in conjunction with the center beam of the multibeam receiver centered on 1390 MHz. Following standard pulse timing procedures, folded profiles were cross-correlated with a template profile to determine times of arrival (TOAs). We used the standard pulsar timing package TEMPO[1], along with the Jet Propulsion Laboratory's DE405 ephemeris for all timing analysis. TOA uncertainties for each pulsar were scaled to achieve reduced $\chi^2 \simeq 1$ in order to improve the estimation of parameter uncertainties. Timing solutions for these pulsars are given in Table 2.1 and derived parameters are given in Table 2.2. Average pulse profiles are shown in Figure 3.1, and timing residuals relative to the models in Table 2.1 are shown in Figures 2.5 and 2.6.

[1]http://pulsar.princeton.edu/tempo

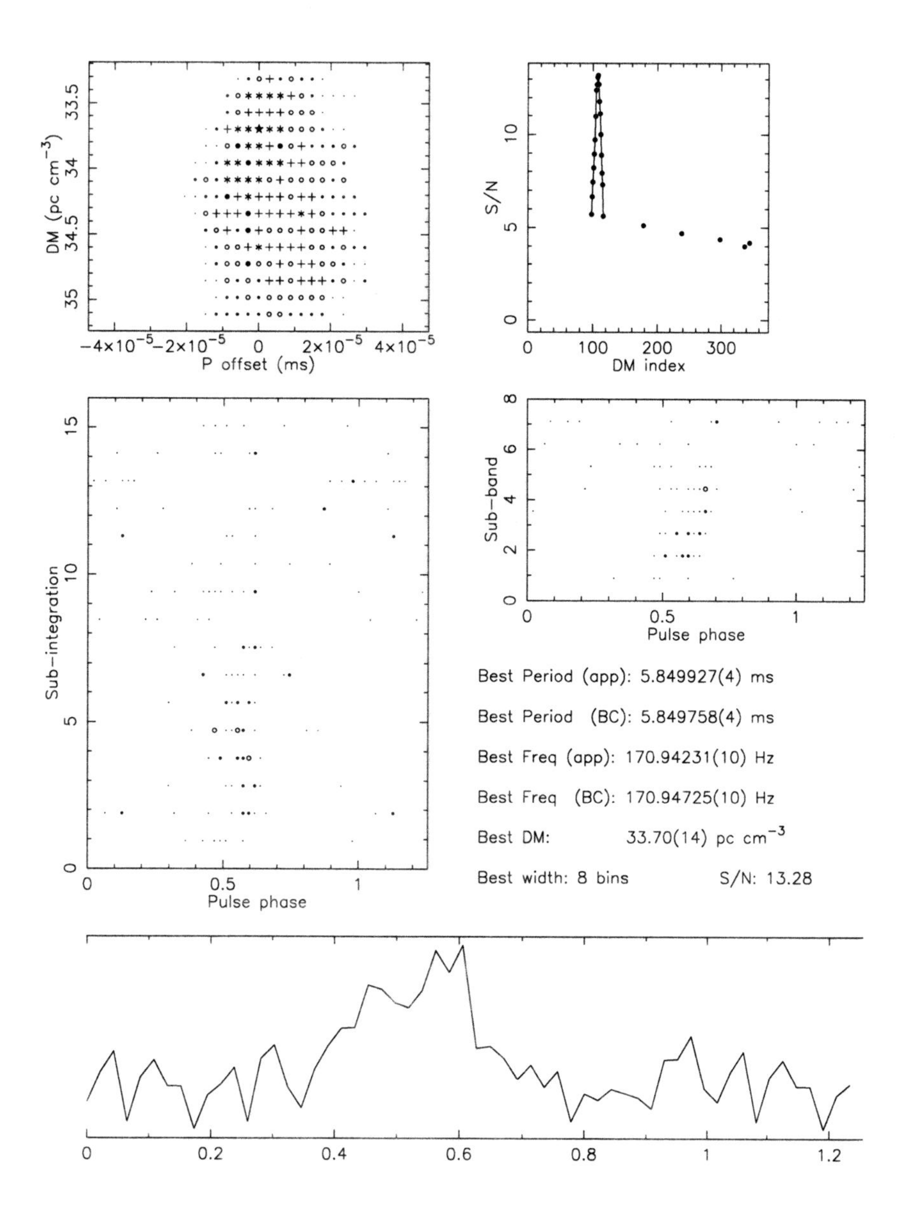

Figure 2.2: Example millisecond pulsar candidate. Shown clockwise from top left are the optimized period and dispersion measure relative to the values found in the frequency domain search, signal-to-noise ratio as a function of dispersion measure trial, folded pulse profile in each of several frequency sub-bands, best folded pulse profile, and folded pulse profile in each of several time sub-integrations. This candidate was confirmed as PSR J1738+0333.

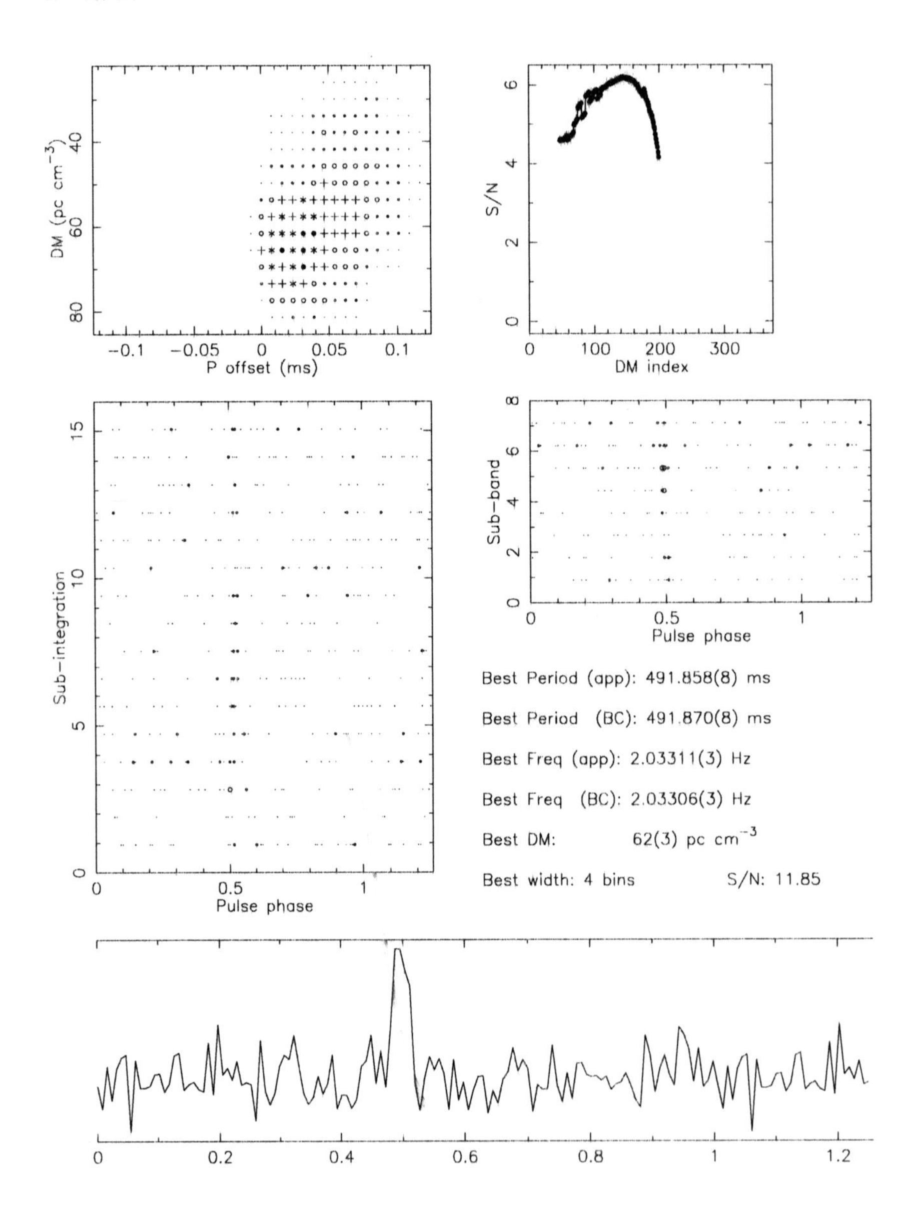

Figure 2.3: Example slow pulsar candidate. Shown clockwise from top left are the optimized period and dispersion measure relative to the values found in the frequency domain search, signal-to-noise ratio as a function of dispersion measure trial, folded pulse profile in each of several frequency sub-bands, best folded pulse profile, and folded pulse profile in each of several time sub-integrations. This candidate was confirmed as PSR J1946−1312.

Table 2.1. Timing model parameters for 19 new slow pulsars

Pulsar	Parameter[a]					
	α (J2000)	δ (J2000)	P (s)	P Epoch (MJD)	$\dot{P}$ 10^{-15}	DM[b] (pc cm^{-3})
J0656−5449	$06^h56^m48^s.990(7)$	$-54^\circ49'14''.92(4)$	0.183156898795(2)	53000.0	0.03191(9)	67.5(10)
J0709−5923	$07^h09^m32^s.533(8)$	$-59^\circ23'55''.60(4)$	0.485268383925(5)	53000.0	0.1260(2)	65(2)
J1231−4609	$12^h31^m45^s.76(14)$	$-46^\circ09'45''.2(3)$	0.87723907778(4)	53000.0	0.0380(17)	76(7)
J1308−4650	$13^h18^m44^s.589(19)$	$-46^\circ50'29''.7(4)$	1.05883304424(3)	53000.0	0.5259(16)	66(10)
J1333−4449	$13^h33^m44^s.829(5)$	$-44^\circ49'26''.22(10)$	0.345602948594(3)	53000.0	0.00054(19)	44.3(17)
J1339−4712	$13^h39^m56^s.5886(18)$	$-47^\circ12'05''.52(3)$	0.1370546579332(4)	53000.0	0.00053(2)	39.9(6)
J1427−4158	$14^h27^m50^s.770(9)$	$-41^\circ58'56''.3(3)$	0.586485556229(18)	53000.0	0.6212(7)	71(3))
J1536−3602	$15^h36^m17^s.382(14)$	$-36^\circ02'58''.8(5)$	1.31975904174(5)	53000.0	0.7900(19)	96(6)
J1609−1930	$16^h09^m05^s.35(12)$	$-19^\circ30'08(9)''$	1.55791724762(7)	53000.0	0.509(3)	37(7)
J1612−2408	$16^h12^m26^s.06(3)$	$-24^\circ08'04(2)''$	0.92383371069(3)	53000.0	1.5736(12)	49(4)
J1635−1511	$16^h35^m47^s.36(4)$	$-15^\circ11'52(3)''$	1.17938703902(8)	53000.0	0.232(4)	54(8)
J1651−7642	$16^h51^m07^s.87(16)$	$-76^\circ42'39''.5(7)$	1.75531107981(18)	53000.0	1.363(8)	80(10)
J1652−1400	$16^h52^m16^s.677(7)$	$-14^\circ00'27''.4(4)$	0.305447058241(3)	53000.0	0.01758(15)	49.5(13)
J1714−1054	$17^h14^m40^s.122(5)$	$-10^\circ54'10''.9(3)$	0.696278743075(9)	53000.0	0.0588(4)	51(3)
J1739+0612	$17^h39^m17^s.966(4)$	$+06^\circ12'28''.4(10)$	0.234169035616(3)	53000.0	0.15640(12)	101.5(13)
J1816−5643	$18^h16^m36^s.464(7)$	$-56^\circ43'42''.10(6)$	0.2179228818474(13)	53000.0	0.00193(6)	52.4(11)
J1841−7845[c]	$18^h41^m25^s.9(4)$	$-78^\circ45'15(4)''$	0.3536025341(6)	53150.0	0.52(4)	41(2)
J1841−7845[d]			0.3536025329(4)	53150.0	0.16(7)	...
J1846−7403	$18^h46^m13^s.78(17)$	$-74^\circ03'04(2)''$	4.8788385261(5)	53000.0	6.06(10)	97(20)
J1946−1312	$19^h46^m57^s.829(10)$	$-13^\circ12'36''.4(6)$	0.491865489484(6)	53000.0	1.9866(3)	60(2)

[a]Figures in parenthesis are uncertainties in the last digit quoted. Uncertainties are calculated from twice the formal error produced by TEMPO.

[b]DM determined from discovery or confirmation observation across 288 MHz wide observing band at 1374 MHz

[c]Parameters obtained from fit to data prior to timing event around MJD 53150

[d]Parameters obtained from fit to data after timing event, holding position fixed

Table 2.2. Derived parameters for 19 new slow pulsars

Pulsar	Parameter								
	S/N	w_{50} (ms)	w_{10} (ms)	l (deg)	b (deg)	d[b] (kpc)	$\|z\|$ (kpc)	τ_c (Myr)	B (10^{12} G)
J0656−5449	11.1	5.4	10.2	264.80	−21.14	3.9	1.4	91	0.077
J0709−5923	14.7	5.1	11.5	270.03	−20.90	3.3	1.2	61	0.25
J1231−4609	18.6	44.6	55.5	299.38	+16.57	2.4	0.69	370	0.18
J1308−4650	12.5	54.6	69.9	306.01	+15.93	1.9	0.53	32	0.76
J1333−4449	15.9	2.4	9.7	310.77	+17.40	1.4	0.41	10000	0.013
J1339−4712	21.9	2.5	4.9	311.42	+14.87	1.2	0.31	4100	0.0086
J1427−4158	11.9	19.2	23.9	321.48	+17.39	2.0	0.61	15	0.61
J1536−3602	57.7	84.6	98.0	336.55	+15.84	3.0	0.82	26	1.0
J1609−1930	19.8	14.6	26.8	354.07	+23.18	1.4	0.55	49	0.90
J1612−2408	16.9	20.8	31.8	351.01	+19.45	1.6	0.54	9.3	1.2
J1635−1511	23.6	39.6	262.0	2.06	+21.08	1.9	0.69	81	0.53
J1651−7642	21.9	83.6	104.9	315.15	−19.95	2.9	0.97	20	1.6
J1652−1400	22.9	11.8	24.1	5.60	+18.58	1.7	0.53	280	0.74
J1714−1054	24.7	7.7	77.2	11.49	+15.78	1.6	0.44	190	0.20
J1739+0612	17.2	7.5	16.5	30.26	+18.86	> 17	> 5.6	24	0.19
J1816−5643	12.3	6.6	22.6	337.67	−17.90	1.5	0.46	1800	0.020
J1841−7845[c]	17.6	22.8	41.4	315.46	−26.08	1.4	0.63	11	0.43
J1841−7845[d]								35	0.24
J1846−7403	30.9	98.2	731.4	320.68	−25.65	> 12	> 5.2	13	5.5
J1946−1312	11.9	11.6	18.9	27.08	−18.18	2.2	0.69	3.9	1.0

[a]For pulsars detected in multiple survey beams, S/N of strongest detection

[b]Distance estimated from dispersion measure using model of Cordes & Lazio (2002)

[c]Parameters obtained from fit to data prior to timing event around MJD 53150

[d]Parameters obtained from fit to data after timing event, holding position fixed

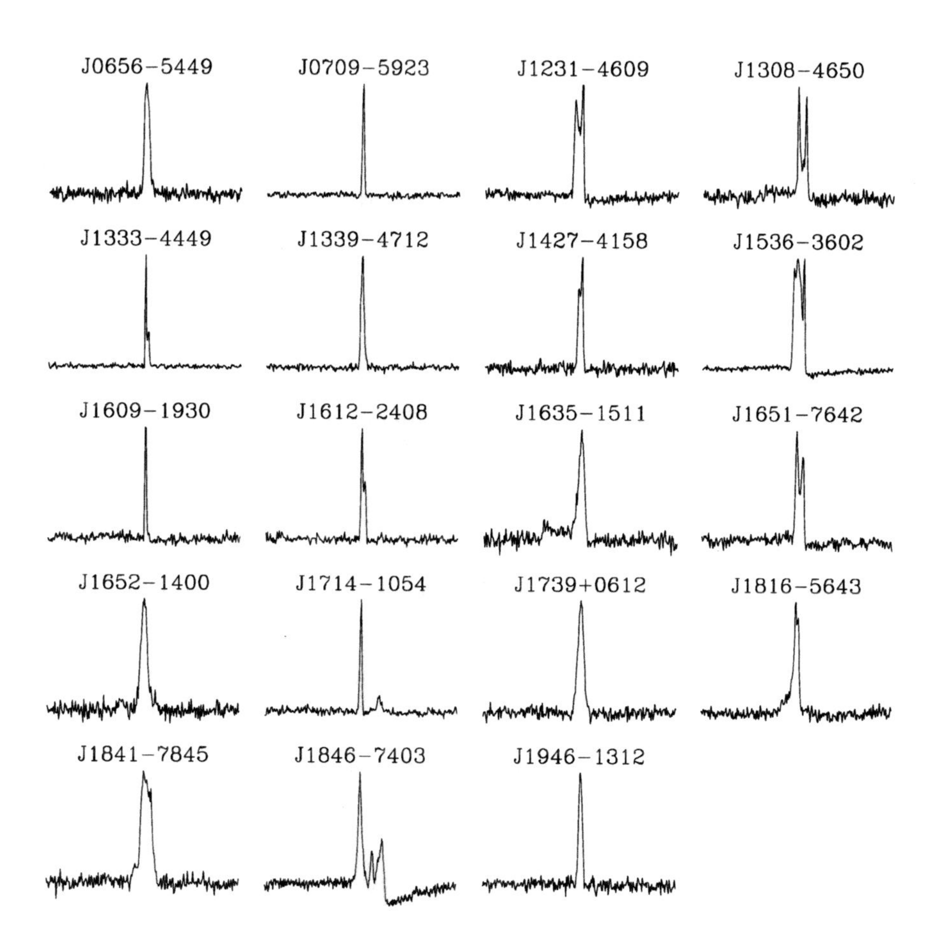

Figure 2.4: Pulse profiles of 19 new slow pulsars.

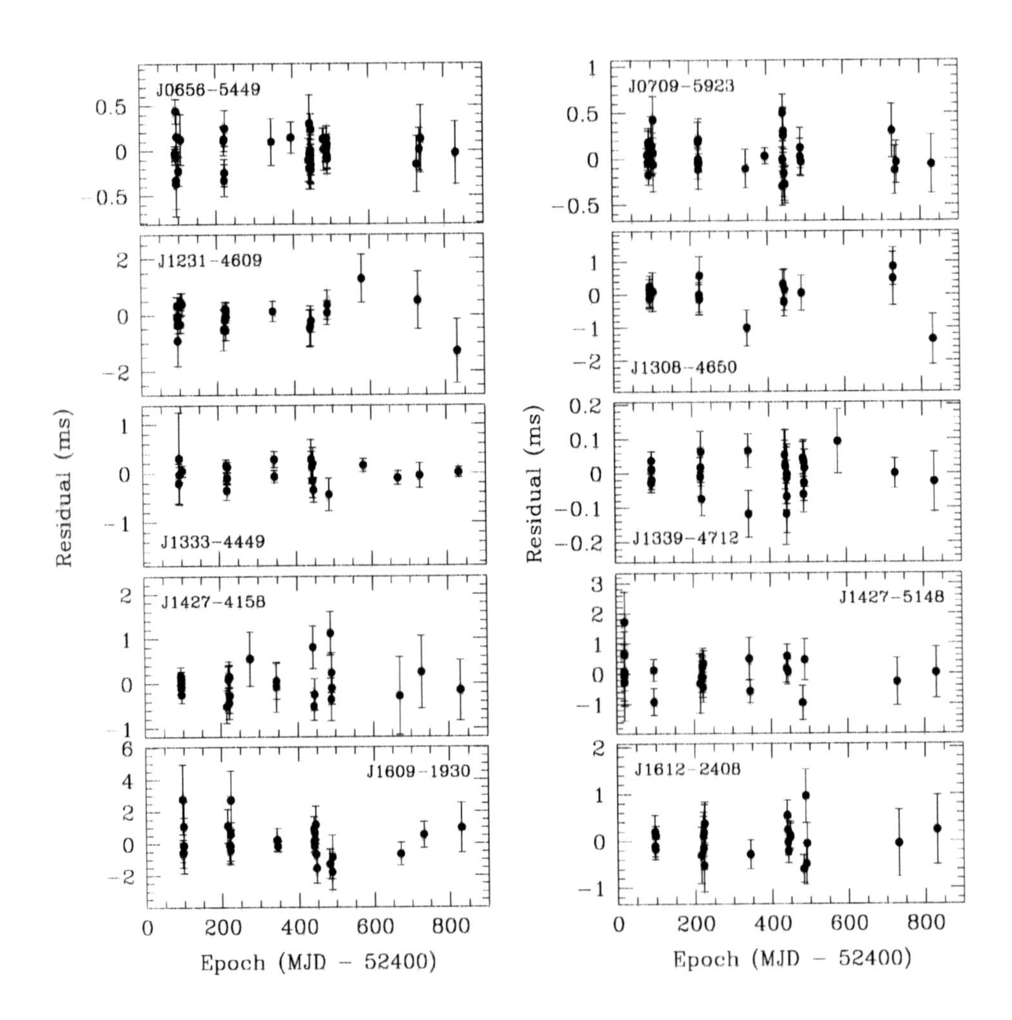

Figure 2.5: Timing residuals for 10 new slow pulsars.

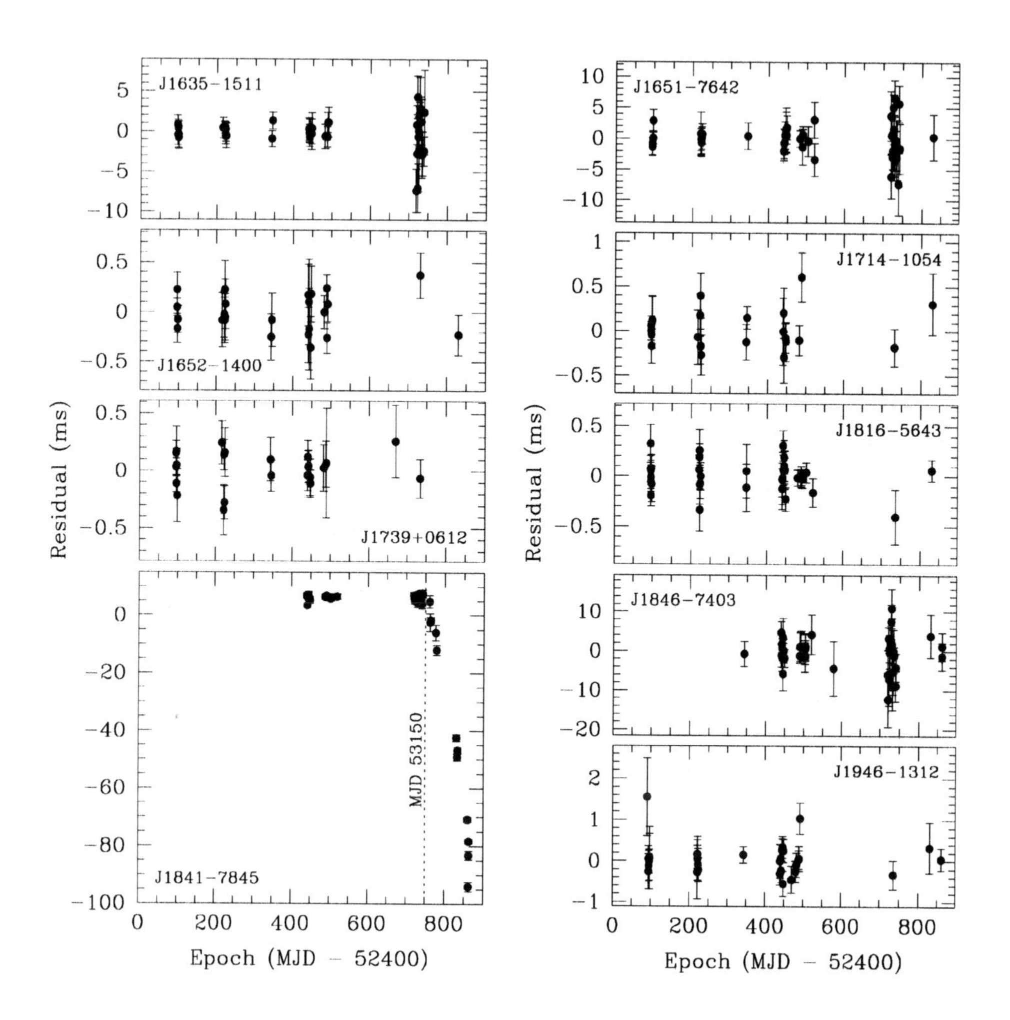

Figure 2.6: Timing residuals for 9 new slow pulsars. The dotted line marks the approximate epoch of a possible glitch-like timing event in PSR J1841−7845.

The timing behavior of these objects is typical of slow pulsars, with the possible exception of PSR J1841−7845. As shown in Figure 2.6, a timing solution which appeared to be valid for nearly a year fails to predict the pulsar phase for recent observations. This timing event, occurring near MJD 53150, does not resemble classical glitch behavior seen in much younger pulsars in which a sudden decrease in P is accompanied by an increase in the spin-down rate, $\dot{P}$ (Shemar & Lyne, 1996). The residuals qualitatively resemble a so-called "slow glitch" as observed in PSR J1825−0935 (Zou et al., 2004) and the fractional change in spin period (-3×10^{-3}) is of similar magnitude; however, our post-event timing solution for J1841−7845 suggests that $\dot{P}$ has decreased to roughly 30% of its pre-event value. We are therefore skeptical that our timing solution is correct for this object, and include it here only as an aid to others who may want to observe this object. Our timing data for this object are sparse; unfortunately several other observations are unusable due to RFI.

Three of these slow pulsars (J1333−4449; J1339−4712; and J1816−5643) have relatively long characteristic ages ($\tau_c = P/2\dot{P} > 10^9\,\mathrm{yr}$) and weak inferred surface dipole magnetic fields ($B_{\mathrm{surf}} = 3.2 \times 10^{19} G s^{-1/2} \sqrt{P\dot{P}} \approx 10^{10}\,\mathrm{G}$) compared to other slow pulsars. In Figure 2.7, we plot B_{surf} vs. P for known pulsars with measured $\dot{P}$ and indicate the objects found in this survey. These three objects fall in the transition region in P – B_{surf} space between the predominantly isolated slow pulsars and the recycled pulsars found most often in binary systems.

In addition to the 26 pulsars discovered in this survey, 36 previously known pulsars were detected (Tab. 2.3). Of the 28 previously known pulsars we failed to detect (Tab. 2.4), only two have published flux densities at 1400 MHz (S_{1400}) greater than 0.4 mJy.

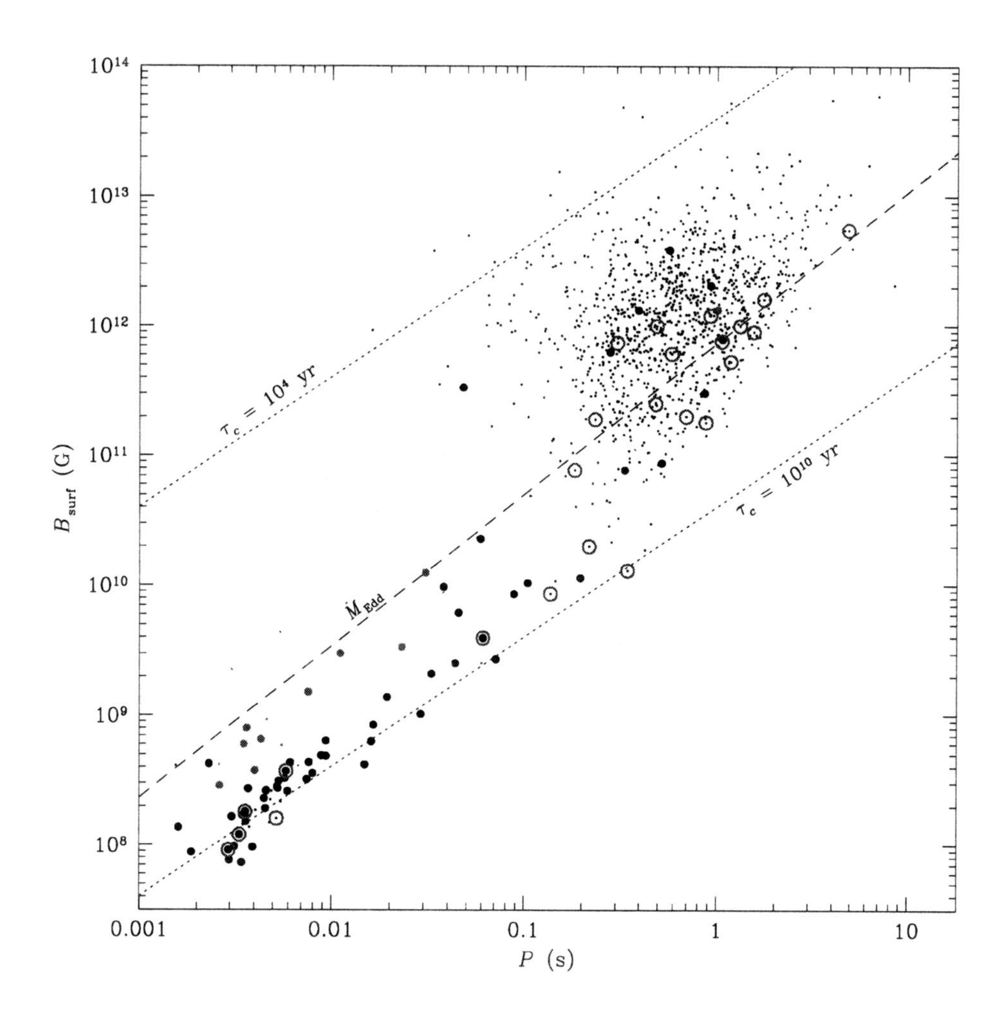

Figure 2.7: Spin period – magnetic field strength diagram. The dashed line shows the equilibrium spin-up line resulting from Eddington accretion. Dotted lines indicate characteristic ages of 10^4 and 10^{10} years. Small points mark isolated objects, filled circles show pulsars in binary systems. Red symbols indicate objects in globular clusters. Blue open circles mark pulsars discovered in this survey, except for J1741+13 (P = 3.75 ms; see Chapter 3) and J1841−7845 (P = 354 ms) for which we do not yet have secure measurements of $\dot{P}$ and hence $B_{\rm surf}$. Where possible, $\dot{P}$ has been corrected for the apparent acceleration induced by proper motion (Shklovskii, 1970).

Table 2.3. Previously known pulsars detected by survey

Pulsar	Parameter[a]					
	P (s)	DM (pc cm^{-3})	S_{1400}[b] (mJy)	l (deg)	b (deg)	S/N[c]
J0711−6830	0.0055	18.4	3.4(5)	279.60	−23.27	14.4
J1034−3224	1.1506	50.8	4.7	272.12	+22.13	87.5
B1114−41	0.9432	40.5	3	284.52	+18.08	93.1
J1141−3107	0.5384	30.8	···	285.82	+29.41	19.2
J1141−3322	0.2915	46.5	1	286.66	+27.29	76.1
J1159−7910	0.5251	59.2	···	300.48	−16.54	31.4
B1237−41	0.5122	44.1	···	300.76	+21.42	36.0
J1320−3512	0.4585	16.4	···	309.62	+27.31	63.7
B1325−43	0.5327	42.0	2	309.95	+18.43	74.9
J1335−3642	0.3992	41.7	···	312.79	+25.34	11.1
J1418−3921	1.0968	60.5	1.4	320.90	+20.47	18.4
B1552−23	0.5326	51.9	0.9(1)	348.52	+22.50	33.5
B1552−31	0.5181	73.0	1.4(4)	342.78	+16.76	64.6
B1600−27	0.7783	46.2	1.7(3)	347.20	+18.77	76.9
J1603−2531	0.2831	53.8	···	348.45	+19.99	225.2
B1620−09	1.2764	68.2	0.6(1)	5.38	+27.18	34.4
B1620−26[d]	0.0111	62.9	2.0(3)	351.05	+15.96	22.2
J1643−1224	0.0046	62.4	3.3(1)	5.75	+21.22	62.9
B1641−68	1.7856	43.0	2	321.91	−14.82	28.1
B1642−03	0.3877	35.7	21.0(6)	14.19	+26.06	443.3
B1701−75	1.1910	37.0	···	316.75	−20.21	29.9
J1713+0747[e]	0.0046	16.0	3	28.83	+25.21	10.1
B1718−02	0.4777	67.0	1.0(2)	20.21	+18.93	15.6
B1726−00	0.3860	41.1	···	23.11	+18.28	20.1
J1736+05	0.9992	42.0	···	29.67	+19.20	205.0
B1737+13	0.8031	48.7	3.9(5)	37.16	+21.67	86.8
J1740+1000	0.1541	23.9	9.2(4)	34.09	+20.26	11.6
B1806−53	0.2610	45.0	···	340.36	−15.90	97.4
B1828−60	1.8894	35.0	···	334.89	−21.19	36.3
J1932−3655	0.5714	59.9	···	2.14	−23.55	31.2
B1937−26	0.4029	50.0	3(1)	13.97	−21.83	40.3
B1940−12	0.9724	28.9	0.7(2)	27.32	−17.16	31.1
B1941−17	0.8412	56.3	0.2	22.38	−19.43	15.1
B1943−29	0.9594	44.3	0.8(3)	11.17	−24.12	20.6

Table 2.3

Pulsar	Parameter[a]					
	P (s)	DM (pc cm^{-3})	S_{1400}[b] (mJy)	l (deg)	b (deg)	S/N[c]
B2003−08	0.5809	32.4	2.8(7)	34.17	−20.31	122.7
B2043−04	1.5469	35.8	1.7(5)	42.74	−27.40	50.6

[a]Parameter values except for detected S/N obtained from http://www.atnf.csiro.au/research/pulsar/psrcat/

[b]Figures in parenthesis are uncertainty in the last digit quoted, where known

[c]For pulsars detected in multiple survey beams, S/N of strongest detection

[d]In the globular cluster M4

[e]Discovered at sub-harmonic of actual spin frequency

The fraction of binary and recycled pulsars discovered in this survey (7 out of 26 new pulsars) is significantly higher than that of the pulsar population as a whole, demonstrating the efficacy of a shallow, high-frequency survey away from the galactic plane for finding MSPs. Figure 2.8 shows the period distribution of newly-discovered pulsars with previously known pulsars in the survey region, excluding those in globular clusters. The distribution of new pulsar periods is clearly weighted toward short periods relative to the previously known pulsars even though MSPs are thought to have slightly steeper spectra than slow pulsars (Toscano et al., 1998), suggesting that the improved effective time resolution afforded by observing at high frequency is a significant advantage compared with the sensitivity of the Parkes Southern Pulsar Survey at 436 MHz (Manchester et al., 1996). The DM distribution of newly discovered pulsars is similar to that of the previously known pulsars (Fig. 2.9). Figure 2.10 shows the galactic distribution of newly discovered pulsars. It is curious that while most of the new pulsars are in the northern side of the galaxy, the distribution of new pulsars in the south is relatively isotropic in the survey region while in the north, it falls off markedly with increasing distance from the galactic plane. We note that a significant part of our survey region, and in fact, three of our newly discovered pulsars are visible with the 305-m Arecibo telescope, further supporting the claim that re-surveying the sky can be profitable.

Table 2.4. Undetected previously known pulsars in survey region

Pulsar	Parameter[a]				
	P (s)	DM ($\mathrm{pc\,cm^{-3}}$)	S_{1400}[b] (mJy)	l (deg)	b (deg)
B0559−57	2.2614	30.0	...	266.56	−29.33
B0904−74	0.5496	51.1	...	289.81	−18.31
B0909−71	1.3629	54.3	...	287.80	−16.24
B1010−23	2.5179	22.5	...	262.20	+26.39
J1047−3032	0.3303	52.5	0.18	273.57	+25.14
B1056−78	1.3474	51.0	...	297.64	−17.56
B1118−79	2.2806	27.4	...	298.77	−17.48
J1455−3330	0.0080	13.6	1.2(1)	330.80	+22.57
B1607−13	1.0184	49.1	...	359.51	+26.95
B1612−29	2.4776	44.8	...	347.46	+15.06
J1650−1654	1.7496	43.2	1.6	2.93	+17.23
B1648−17	0.9734	33.5	0.3(1)	2.89	+16.88
B1657−13	0.6410	60.4	...	7.58	+17.59
J1720+2150	1.6157	41.1	...	44.03	+29.36
B1732−02	0.8394	65.0	...	21.98	+15.92
J1752+2359	0.4091	36.0	...	49.17	+23.06
J1756+18	0.7440	77.0	...	43.82	+20.21
J1813+1822	0.3364	60.8	...	45.62	+16.39
B1851−79	1.2792	39.0	...	314.39	−27.05
J1910−5959A[c]	0.0033	33.7	0.22	336.59	−25.73
J1910−5959B[c]	0.0084	33.3	0.06	336.56	−25.62
J1910−5959C[c]	0.0053	33.2	0.30	336.53	−25.66
J1910−5959D[c]	0.0090	33.3	0.07	336.56	−25.62
J1910−5959E[c]	0.0046	33.3	0.09	336.56	−25.62
J1940−2403	1.8553	63.3	...	15.91	−21.01
J1947−4215	1.7981	35.0	...	357.25	−27.71
B1946−25	0.9576	23.1	0.4(1)	15.33	−23.39
J2005−0020	2.2797	35.9	0.4	41.40	−16.64

[a]Parameter values obtained from http://www.atnf.csiro.au/research/pulsar/psrcat/

[b]Figures in parenthesis are uncertainty in the last digit quoted, where known.

[c]In the globular cluster NGC 6752

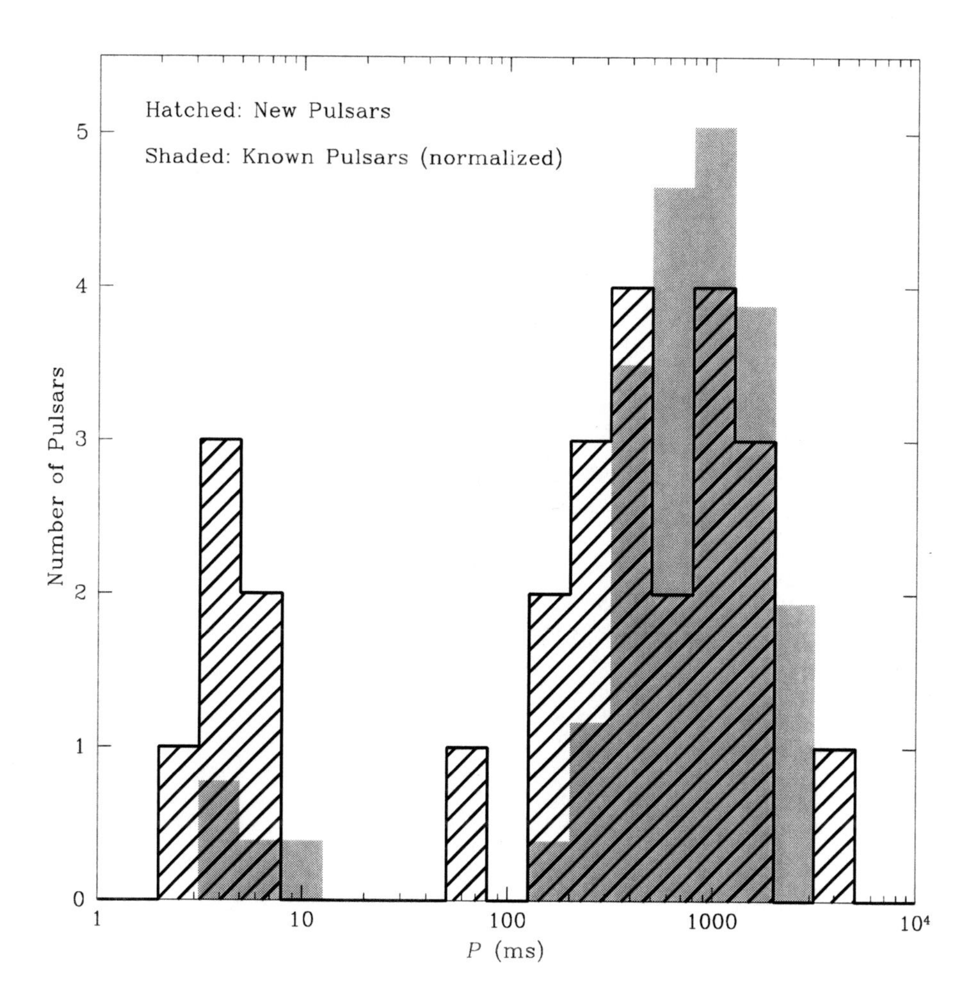

Figure 2.8: Spin period distribution of pulsars in survey region. The hatched histogram shows the distribution of the 26 newly discovered pulsars, while the grey shaded histogram shows the 57 previously known field pulsars in the survey region, normalized to have the same total area as the new pulsars.

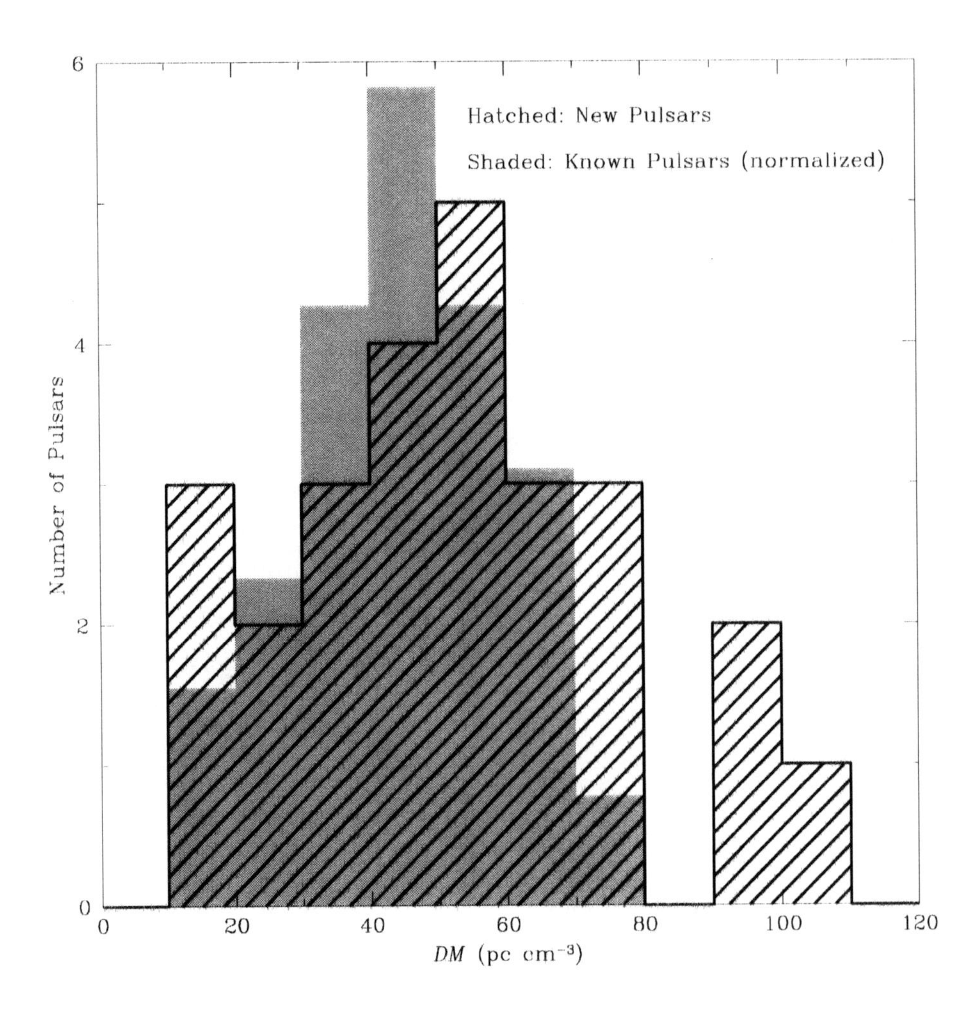

Figure 2.9: Dispersion measure distribution of pulsars in survey region. The hatched histogram shows the distribution of the 26 newly discovered pulsars, while the grey shaded histogram shows the 57 previously known field pulsars in the survey region, normalized to have the same total area as the new pulsars.

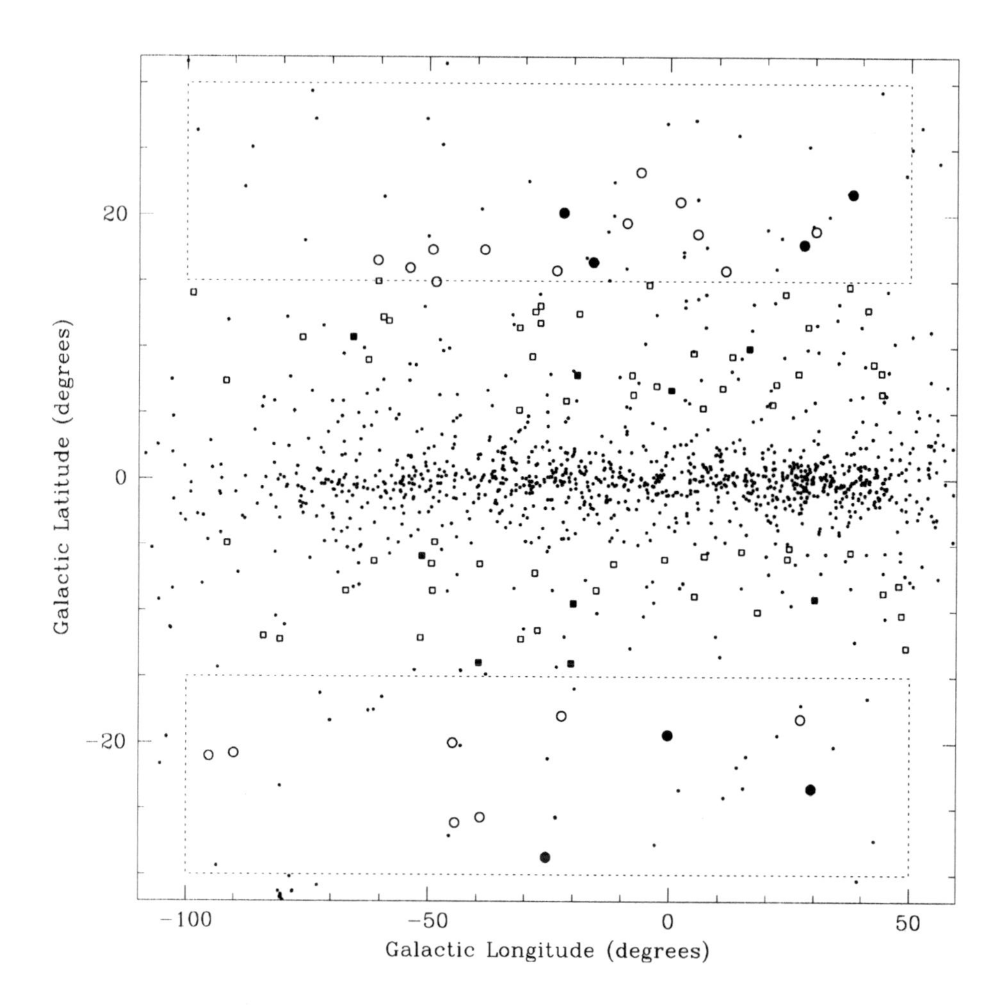

Figure 2.10: Galactic distribution of pulsars. The survey region is denoted by the dashed boxes, and newly discovered pulsars are shown as circles in these regions. Open circles represent the 19 slow pulsars reported here, while filled circles denote the 7 recycled pulsars described in Chapters 3 and 4. Open and filled squares, respectively, represent the slow and recycled pulsars discovered in the previous Swinburne Intermediate Latitude Pulsar Survey (Edwards et al., 2001), and other known pulsars are shown as points.

2.4 Implications for Sub-Millisecond Pulsars

It is curious that the fastest spinning neutron star known, PSR B1937+21 with $P = 1.56$ ms, was the first millisecond pulsar discovered (Backer et al., 1982). This exceptional object was discovered in a targeted search and until fairly recently, few large-area pulsar surveys were sensitive to pulsars with spin periods of order a millisecond. It is interesting to ask whether the lack of observed pulsars with periods faster than that of B1937+21 can be explained by observational selection effects, or if it requires these putative "sub-millisecond" pulsars to be an extremely rare population.

If pulsars with spin periods shorter than 1.56 ms do not exist, it may be simply because the recycling process — the mechanism by which a low-mass star evolves off the main sequence and donates mass and angular momentum to its neutron star companion — does not efficiently produce such rapidly rotating objects, or once produced, their rotation quickly slows because of gravitational wave emission (Bildsten, 2003). However, it may be that the minimum spin period of neutron stars is imposed by centripetal acceleration, with the stars breaking apart when the rotational speed at the equator exceeds the gravitational escape speed.

This question has important implications for the equation of state of dense matter which relates the mass and radius of neutron stars. Figure 2.11 shows the mass-radius relationship for a number of possible neutron star equations of state (EOSs), as well as the relationship between breakup spin period and radius for several relevant values of neutron star mass. Given the observed spin period and mass distribution of radio pulsars, the constraints on the EOS are not very strong: if PSR B1937+21 has the same mass as PSR J1909−3744, its radius can still be as large as 16 km. Progress on this front is most likely to come from constraints on the minimum neutron spin period, either from discovery of faster-rotating pulsars or a better understanding of the underlying population.

Through a detailed analysis of the known pulsar population, Cordes & Chernoff (1997) found that the most likely minimum period of the underlying period distribution is not much less than that of PSR B1937+21, but that at the 95% confidence level, spin periods as short as 1 ms could not be ruled out, and at the 99% confidence level, extended as low as 0.65 ms. More recently, spin periods for 11 neutron stars obtained through X-ray observations whose sensitivity does not suffer as badly shortward of ~ 1 ms as typical radio surveys have been

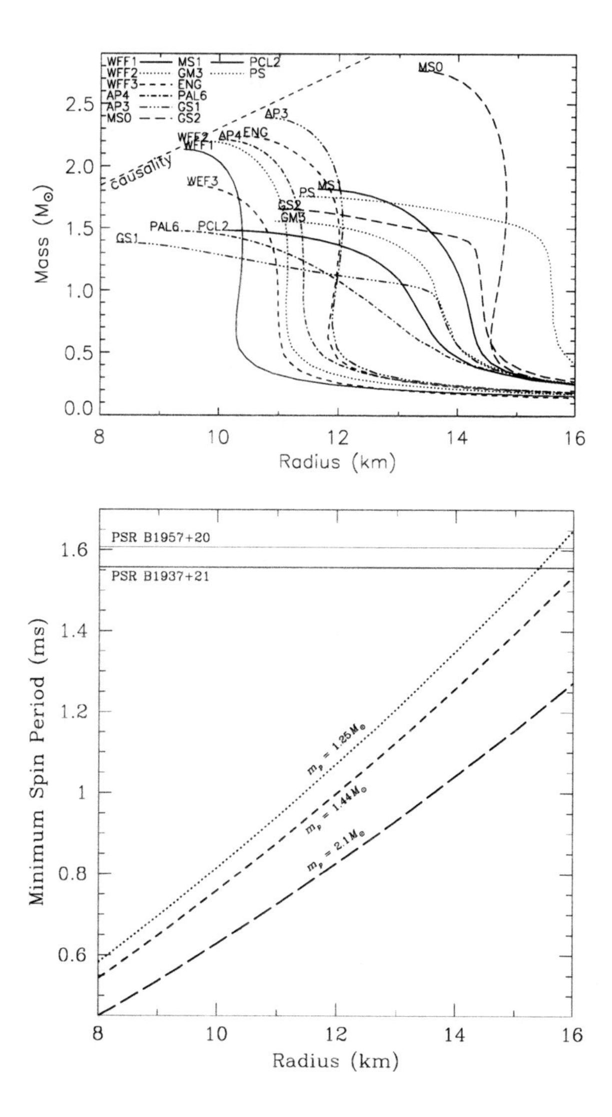

Figure 2.11: Neutron star mass, radius, and spin period. Top: mass vs. radius for a variety of possible equations of state, from Lattimer & Prakash (2001). Bottom: minimum (breakup) spin period vs. radius for example neutron stars of mass $1.25\,M_{\odot}$ (slow pulsar J0737−3039B, Lyne et al., 2004), $1.44\,M_{\odot}$ (millisecond pulsar J1909−3744, Chapter 5), and $2.1\,M_{\odot}$ (millisecond pulsar J0751+1807, Nice et al., 2004), with horizontal lines indicating the two shortest known pulsar spin periods.

used to infer a minimum spin period of 1.3 ms to 95% confidence, assuming a uniform distribution down to a cutoff period (Chakrabarty et al., 2003), and through application of a similar analysis to the radio pulsar population, McLaughlin et al. (2004) find a minimum spin period of 1.2 ms at 95% confidence. As an exercise, we have attempted a simple analysis of the millisecond radio pulsar population which includes the results of several pulsar surveys in roughly the last decade. We consider the known population of pulsars with $P < 10$ ms, and ask the following question: if each of these pulsars could be made to spin faster without changing any of its other parameters (e.g., duty cycle, DM, and luminosity), what is the shortest spin period at which the discovery survey could still have detected it? We consider only pulsars found in modern surveys with good sensitivity to MSPs, excluding any objects found in targeted searches (globular clusters, known steep spectrum objects, γ-ray sources, etc.), or those for which we cannot reliably estimate parameters impacting detection sensitivity. This approach to understanding the underlying population of pulsars has the attractive feature that it takes selection effects very naturally into account; however, it is simplistic and assumes a homogeneity in the population that is far from assured. We have also neglected the increasing difficulty posed by a given orbital acceleration as the pulsar period decreases. Furthermore, very short period pulsars may preferentially exist in short-period binary systems (Burderi et al., 2001), and the computational power and techniques to efficiently detect such systems has only recently become available (Ransom, 2001).

In Figure 2.12, we plot this minimum detection period $P_{\rm min}$, calculated by Equation 2.1, and taking into account subtleties such as the size of the DM steps and number of harmonics summed in the search against the actual spin period of 36 millisecond pulsar. Of these, we find that 20 could have been detected at spin periods shorter than 1.56 ms. Taken at face value, this implies that pulsars with spin periods between 0.30 ms (the smallest $P_{\rm min}$ value resulting from this analysis) and 1.56 ms must be about 50 times less common per octave than pulsars with periods between 1.56 ms and 10 ms. However, we also find that fully one fourth of the pulsars we consider should not have been detected at all by the surveys that discovered them! This result is explained by interstellar scintillation, which can cause a pulsar's observed flux to deviate widely from the average value which we have used where possible in our calculations. Taken together with the likelihood that formal sensitivity calculations for radio pulsar surveys may be overly optimistic at short periods

(Edwards et al., 2001), we believe this analysis cautions against excluding the possibility of pulsars with periods shorter than 1.56 ms based on the sensitivities of pulsar surveys to date. However, it does appear that very short period pulsars, if they exist, are uncommon and are likely to be found only through intensive search efforts with very high time resolution.

2.5 Conclusions

We have completed a survey for pulsars covering 10% of the sky with the 21-cm multibeam receiver system at the 64-m Parkes radio telescope. This survey was very successful, discovering 7 new recycled pulsars (described in detail in Chapters 3 and 4) and 19 slow pulsars, and re-detecting 36 previously known pulsars in the survey region.

Taken together with the previous Swinburne Intermediate Latitude Survey which discovered 69 new pulsars including 8 recycled pulsars, we now have a consistent census of the pulsar population over $\sim 7,100$ square degrees between 5° and 30° from the galactic plane, which will be very valuable for modeling the underlying population.

We thank W. van Straten for help with survey observations, and the Parkes Multibeam Pulsar Survey collaboration for making the data acquisition hardware and software used for this survey available to the community. BAJ and SRK thank NSF and NASA for supporting their research. The Parkes telescope is part of the Australia Telescope which is funded by the Commonwealth of Australia for operation as a National Facility managed by CSIRO.

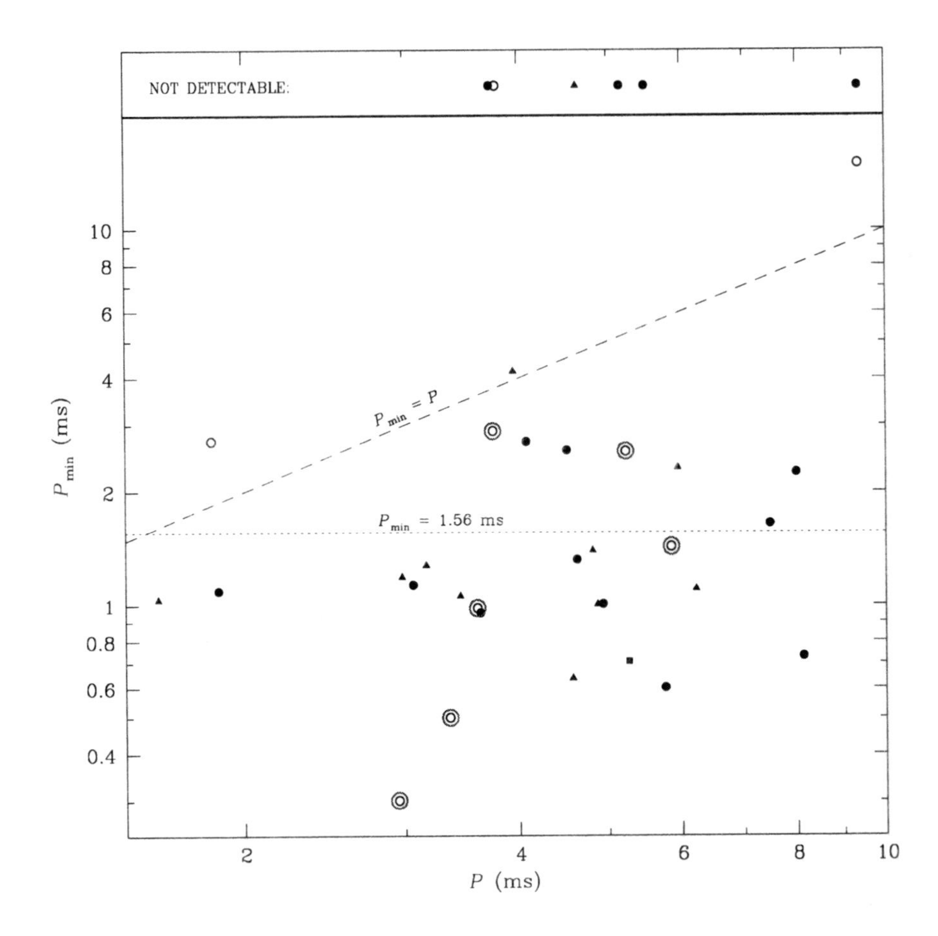

Figure 2.12: Minimum detectable spin period vs. spin period for 36 MSPs. The minimum spin period at which a given millisecond pulsar could have been detected given the sensitivity of the discovery survey and pulsar's DM and pulse shape is plotted against observed spin period. Open circles show pulsars discovered in surveys with the 21-cm Parkes Multibeam system (This work; Manchester et al., 2001; Hobbs et al., 2004), filled circles are pulsars from the Parkes 70 cm pulsar survey (Manchester et al., 1996; Lyne et al., 1998), filled triangles are pulsars from Arecibo 430 MHz surveys (Camilo et al., 1993; Nice et al., 1993; Foster et al., 1995; Nice et al., 1995; Camilo et al., 1996; Ray et al., 1996; Lommen et al., 2000), and filled squares are from the Jodrell Bank 600 MHz survey (Nicastro et al., 1995). Blue circles indicate pulsars discovered in this survey, described fully in Chapters 3 and 4. In calculating P_{min}, average flux densities have been used where available; otherwise, the calculation was based on the discovery signal-to-noise ratio. The horizontal dotted line indicates the period of the fastest spinning known pulsar, PSR B1937+21. Pulsars above the slanted dashed $P_{min} = P$ line should not have been detected in the discovery survey, but could have been detected at the plotted value of P_{min}; pulsars above the solid horizontal line are not detectable in the discovery survey at any period. These pulsars, which according to this calculation should not have been detected, simply indicate that the pulsars' flux density during the survey observation was greater than the published average flux density.

Chapter 3

Discovery of Six Recycled Pulsars in a High Galactic Latitude Survey†

B. A. Jacoby[a], M. Bailes[b], S. Ord[b], H. Knight[b,c], A. Hotan[b,c], and M. H. van Kerkwijk[d]

[a]Department of Astronomy, California Institute of Technology, MS 105-24, Pasadena, CA 91125; baj@astro.caltech.edu.

[b]Centre for Astrophysics and Supercomputing, Swinburne University of Technology, P.O. Box 218, Hawthorn, VIC 31122, Australia; mbailes@swin.edu.au, sord@swin.edu.au, ahotan@swin.edu.au.

[c]Australia Telescope National Facility, CSIRO, P.O. Box 76, Epping, NSW 1710, Australia.

[d]Department of Astronomy and Astrophysics, University of Toronto, 60 St. George Street, Toronto, ON M5S 3H8, Canada; mhvk@astro.utoronto.ca.

Abstract

We present six recycled pulsars discovered during a 21-cm survey of approximately 4,150 deg^2 between 15° and 30° from the galactic plane using the Parkes radio telescope. One new pulsar, PSR J1528−3146, has a relatively long spin period and a massive white dwarf companion. The five remaining pulsars have short spin periods ($P < 10\,\mathrm{ms}$); four of these have white dwarf binary companions and one is isolated.

†Part of a manuscript in preparation for publication in *The Astrophysical Journal*

3.1 Introduction

Because most pulsars are thought to descend from massive stars in the disk of the galaxy, the bulk of pulsar search efforts have historically been concentrated near the galactic plane. However, the distribution of detectable recycled pulsars is expected to be relatively isotropic because compared to young pulsars, they have had a longer lifetime to migrate away from their birthplace in the galactic disk and their shorter periods strongly limit the distance to which they can be detected in the electron-rich galactic plane. These facts, combined with the relative insensitivity to dispersion and lower sky temperatures afforded by high-frequency observations, suggested that a 21 cm survey for pulsars away from the galactic plane would be extremely productive. The success of the Swinburne Intermediate Latitude Pulsar Survey demonstrated the validity of this approach, discovering eight recycled pulsars in $\sim 3,000\,\mathrm{deg}^2$ at galactic latitudes between 5° and 15° (Edwards & Bailes, 2001a,b; Edwards et al., 2001).

3.2 A High Latitude Pulsar Survey

We carried out a pulsar search using the 13-beam multibeam receiver on the 64 m Parkes radio telescope from January 2001 to December 2002. This survey covered $\sim 4,150\,\mathrm{deg}^2$ in the region $-100° < l < 50°$, $15° < |b| < 30°$. Relatively short 265 s integrations gave a sensitivity which is well-matched to the expected scale height and luminosity distribution of the pulsar population and allowed us to complete the 7,232 survey pointings in about four weeks of observing. The signals from each beam were processed and digitized by a $2 \times 96 \times 3$ MHz filterbank operating at a center sky frequency of 1374 MHz and one-bit sampled every 125 μs, providing good sensitivity to fast pulsars with low to moderate dispersion measures. This observing methodology is identical to that employed for the Swinburne Intermediate Latitude and differs from that of the Parkes Multibeam Pulsar Survey (Manchester et al., 2001) only in sampling period (125 μs vs. 250 μs) and integration time (265 s vs. 2100 s).

The resulting 2.4 TB of data were searched for pulsar-like signals using standard techniques with the 64 Compaq Alpha workstations at the Swinburne Centre for Astrophysics and Supercomputing, resulting in the discovery of 26 new pulsars (Chapter 2). Of these 26 new pulsars, seven are recycled. One of them, PSR J1909−3744, is an exceptionally interesting millisecond pulsar and has been reported elsewhere (Jacoby et al., 2003, see Chapter 4). The other six are described here.

3.3 Discovery and Timing of Six Recycled Pulsars

These six objects belong to the class of recycled pulsars with relatively weak magnetic fields and small spindown rates. Five are "millisecond" pulsars (MSPs) with spin periods well under 10 ms; the sixth has a longer spin period, but is still clearly recycled due to its large characteristic age (τ_c). One of the MSPs is isolated; the other pulsars are in binary systems with white dwarf companions. Average pulse profiles of the six pulsars are shown in Figure 3.1.

We have begun a systematic timing program at Parkes for these and other pulsars discovered in this survey, primarily using the $2 \times 512 \times 0.5$ MHz filterbank at 1390 MHz with occasional observations using the $2 \times 256 \times 0.125$ MHz filterbank at 660 MHz to determine the dispersion measure (DM).

We followed standard pulsar timing procedures: folded pulse profiles from individual observations were cross-correlated with a high signal-to-noise template profile to determine an average pulse time of arrival (TOA) corrected to UTC(NIST). The standard pulsar timing package TEMPO[1], along with the Jet Propulsion Laboratory's DE405 ephemeris, was used for all timing analysis. TOA uncertainties for each pulsar were multiplied by a factor between 1.12 and 1.65 to achieve reduced $\chi^2 \simeq 1$. Two of the new pulsars, PSR 1738+0333 and PSR J1933−6211, have very small orbital eccentricities (e), giving rise to a strong covariance between the time of periastron (T_0) and longitude of periastron (ω) in these systems. For these pulsars, we have used the ELL1 binary model which replaces ω, T_0, and e with the time of ascending node ($T_{\rm asc}$) and the Laplace-Lagrange parameters $e \sin \omega$ and $e \cos \omega$ (Lange et al., 2001). We have used the DD model (Damour & Deruelle, 1985, 1986) for PSR J1528−3146 and PSR J1600−3053.

We have not yet obtained a phase-connected timing solution for the binary MSP PSR J1741+13. For this object, orbital parameters were obtained by fitting a model to observed spin periods obtained from multiple observations with the 96-channel survey filterbank. As our current knowledge of its position is limited by the telescope beam size, the name of this pulsar may

[1] http://pulsar.princeton.edu/tempo

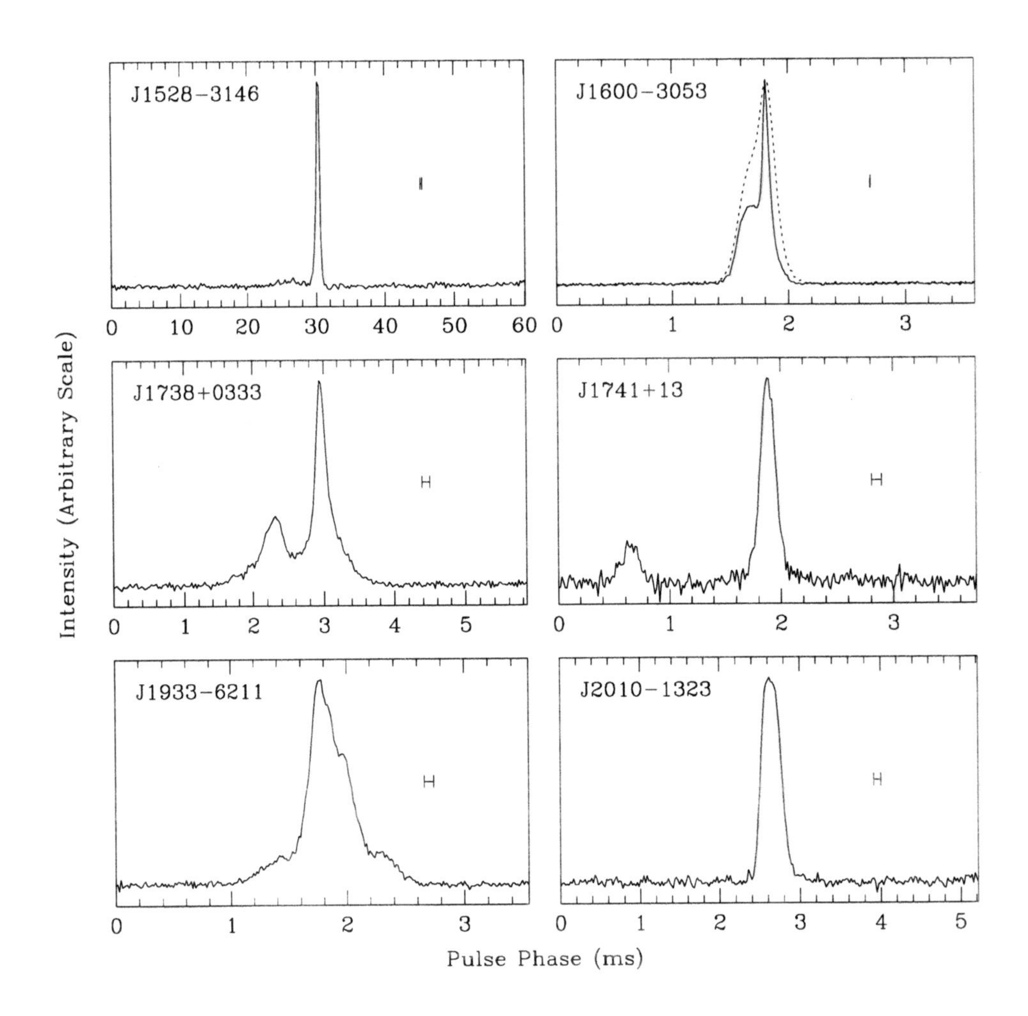

Figure 3.1: Average pulse profiles of six recycled pulsars at 1.4 GHz. PSR J1600-3053 profile measured with CPSR2 (solid) and 512 Parkes filterbank, and J1741+1354 profile measured by 96 channel Parkes filterbank. All others measured by 512 channel Parkes filterbank. Horizontal bars represent the time resolution of the observing system arising from the differential dispersion within a filterbank channel and the sampling interval, except for J1600−3053 where horizontal bar indicates 2 μs time resolution of coherently dedispersed pulse profile.

change when its position is determined more precisely.

Astrometric, spin, binary, and derived parameters for all pulsars are given in Tables 3.1 and 3.2. Timing residuals for the five pulsars with timing solutions are shown in Figure 3.2.

3.3.1 PSR J1528−3146: Massive Companion

PSR J1528−3146 has the longest spin period ($P = 60.8$ ms) of the six new pulsars reported here. The minimum companion mass (obtained from the mass function by assuming an edge-on orbit and a pulsar mass of $1.35\,M_{\odot}$) is $0.94\,M_{\odot}$; the system's circular orbit suggests that the companion must be a CO or ONeMg white dwarf. Only a handful of such intermediate mass binary pulsar (IMBP) systems are known (Edwards & Bailes, 2001b; Camilo et al., 2001). Of all low-eccentricity binary pulsars, this object has the second highest minimum companion mass and fifth highest projected orbital velocity. The orbital parameters of PSR J1528−3146 are broadly similar to those of PSR J1157−5112, the low-eccentricity binary pulsar with the most massive white dwarf companion (Edwards & Bailes, 2001a). We have detected a potential blue optical counterpart at the pulsar timing position (Chapter 6).

3.3.2 PSR J1600−3053: High Timing Precision LMBP

Our timing results obtained with the Parkes filterbank for this 3.6 ms pulsar have a weighted RMS residual of only $1.55\,\mu$s over three years. However, observations with the Caltech-Parkes-Swinburne Recorder II (CPSR2; see Appendix B and Bailes, 2003), a wide-bandwidth coherent dedispersion backend at Parkes, reveal a very narrow profile component which allows for extremely precise temporal localization of the pulsar signal. Initial results with CPSR2 suggest the potential for substantially improved timing precision and will be reported in a future paper. This pulsar will be an important part of pulsar timing array experiments aimed at detecting low-frequency gravitational waves emitted by coalescing supermassive black hole binaries (Jaffe & Backer, 2003).

3.3.3 PSR J1738+0333: Short Binary Period LMBP

Only one white dwarf binary in the galactic disk has an orbital period shorter than that of J1738+0333. While not especially bright, this pulsar has a narrow pulse width and may

Table 3.1. Pulsar Parameters for J1528−3146, J1600−3053, and J1738+0333

Parameter[a]	J1528−3146	J1600−3053	J1738+0333
Right ascension, α_{J2000}	$15^h28^m34^s.9542(2)$	$16^h00^m51^s.90392(2)$	$17^h38^m53^s.96317(5)$
Declination, δ_{J2000}	$-31°46'06''.836(8)$	$-30°53'49''.325(2)$	$+03°33'10''.839(2)$
Proper motion in α, μ_α (mas yr^{-1})	...	-0.91(51)	5.6(17)
Proper motion in δ, μ_δ (mas yr^{-1})	...	-4.0(15)	4(4)
Pulse period, P (ms)	60.82223035146(1)	3.597928452222642(6)	5.8500957646929(6)
Reference epoch (MJD)	52500.0	52500.0	52500.0
Period derivative, $\dot{P}$ (10^{-20})	24.9(1)	0.9479(4)	2.409(4)
Dispersion measure, DM (pc cm^{-3})	18.163(6)	52.333(1)	33.778(9)
Binary model	DD	DD	ELL1
Binary period, P_b (d)	3.180345754(3)	14.348457554(4)	0.3547907344(4)
Projected semimajor axis, $a \sin i$ (lt-s)	11.452324(5)	8.8016571(4)	0.343430(2)
Orbital eccentricity, e	0.000213(1)	0.00017371(8)	...
Longitude of periastron, ω (deg)	296.83635±0.2	181.768043±0.03	...
Time of periastron, T_0	52502.4013744±0.002	52506.3711244±0.001	...
$e \sin \omega$ ($\times 10^{-6}$)	...	...	-0.009(8000)
$e \cos \omega$ ($\times 10^{-6}$)	...	...	4(9)
Time of ascending node, T_{asc} (MJD)	...	...	52500.1938656(2)
Weighted RMS timing residual (μs)	11.1	1.55	3.99
Derived Parameters			
Orbital eccentricity, e	...	...	0.000004(90)
Longitude of periastron, ω (deg)	...	...	359.8793±98
Time of periastron, T_0 (MJD)	...	...	52500.5485374±0.1
Minimum companion mass $m_{c\ min}$ ($M_\odot$)	0.94	0.20	0.09
Galactic longitude, l (deg)	337.94	344.09	27.72
Galactic latitude, b (deg)	20.22	16.45	17.74
DM-derived distance, d (kpc)[b]	0.80	1.53	1.41
Distance from Galactic plane, $\|z\|$ (kpc)	0.28	0.43	0.43
Transverse velocity, $v_\perp$ (km s^{-1})[c]	...	30^{+16}_{-15}	44(23)
Surface magnetic field, B_{surf} (10^8 G)	39.3	1.8[d]	3.7[d]
Characteristic age, τ_c (Gyr)	3.9	6.2[d]	4.0[d]
Pulse FWHM, w_{50} (ms)	0.59	0.079	0.43
Pulse width at 10% peak, w_{10} (ms)	1.29	0.41	1.44
Discovery S/N	28.0	16.7	13.3

[a]Figures in parenthesis are uncertainties in the last digit quoted. Uncertainties are calculated from twice the formal error produced by TEMPO.

[b]From the model of Cordes & Lazio (2002)

[c]Stated uncertainty in transverse velocity is based only on uncertainty in proper motion; the distance is taken as exact.

[d]Corrected for secular acceleration based on measured proper motion and estimated distance (Shklovskii, 1970)

Table 3.2. Pulsar Parameters for J1741+13, J1933−6211, and J2010−1323

Parameter[a]	J1741+13	J1933−6211	J2010−1323
Right ascension, α_{J2000}	$17^h41^m37^s \pm 1^m$	$19^h33^m32^s.4272(3)$	$20^h10^m45^s.9196(2)$
Declination, δ_{J2000}	$+13°54'41'' \pm 14'$	$-62°11'46''.881(4)$	$-13°23'56''.027(6)$
Pulse period, P (ms)	3.7471544(6)	3.543431438847(1)	5.223271015190(1)
Reference epoch (MJD)	···	53000.0	52500.0
Period derivative, $\dot{P}$ (10^{-20})	···	0.37(1)	0.482(7)
Dispersion measure, DM (pc cm^{-3})	24.0(3)	11.499(7)	22.160(2)
Binary model	···	ELL1	···
Binary period, P_b (d)	16.335(2)	12.81940650(4)	···
Projected semimajor axis, $a \sin i$ (lt-s)	11.03(6)	12.281575(3)	···
$e \sin \omega$ ($\times 10^{-6}$)	···	1.1(4)	···
$e \cos \omega$ ($\times 10^{-6}$)	···	-0.55(50)	···
Time of ascending node, T_{asc} (MJD)	52846.22(1)	53000.4951005(5)	···
Weighted RMS timing residual (μs)	···	6.06	4.25
Derived Parameters			
Orbital eccentricity, e	···	0.0000013(4)	···
Longitude of periastron, ω (deg)	···	115.93036±22	···
Time of periastron, T_0 (MJD)	···	53004.6233183±0.8	···
Minimum companion mass $m_{c\ min}$ ($M_\odot$)	0.24	0.32	···
Galactic longitude, l (deg)	37.94	334.43	29.45
Galactic latitude, b (deg)	21.64	-28.63	-23.54
DM-derived distance, d (kpc)[b]	0.91	0.52	1.02
Distance from Galactic plane, $\|z\|$ (kpc)	0.34	0.25	0.41
Surface magnetic field, B_{surf} (10^8 G)	···	1.2	1.6
Characteristic age, τ_c (Gyr)	···	15	17
Pulse FWHM, w_{50} (ms)	0.16	0.36	0.28
Pulse width at 10% peak, w_{10} (ms)	0.32	1.03	0.44
Discovery S/N	10.7	22.9	12.4

[a]Figures in parenthesis are uncertainties in the last digit quoted. Uncertainties are calculated from twice the formal error produced by TEMPO.

[b]From the model of Cordes & Lazio (2002)

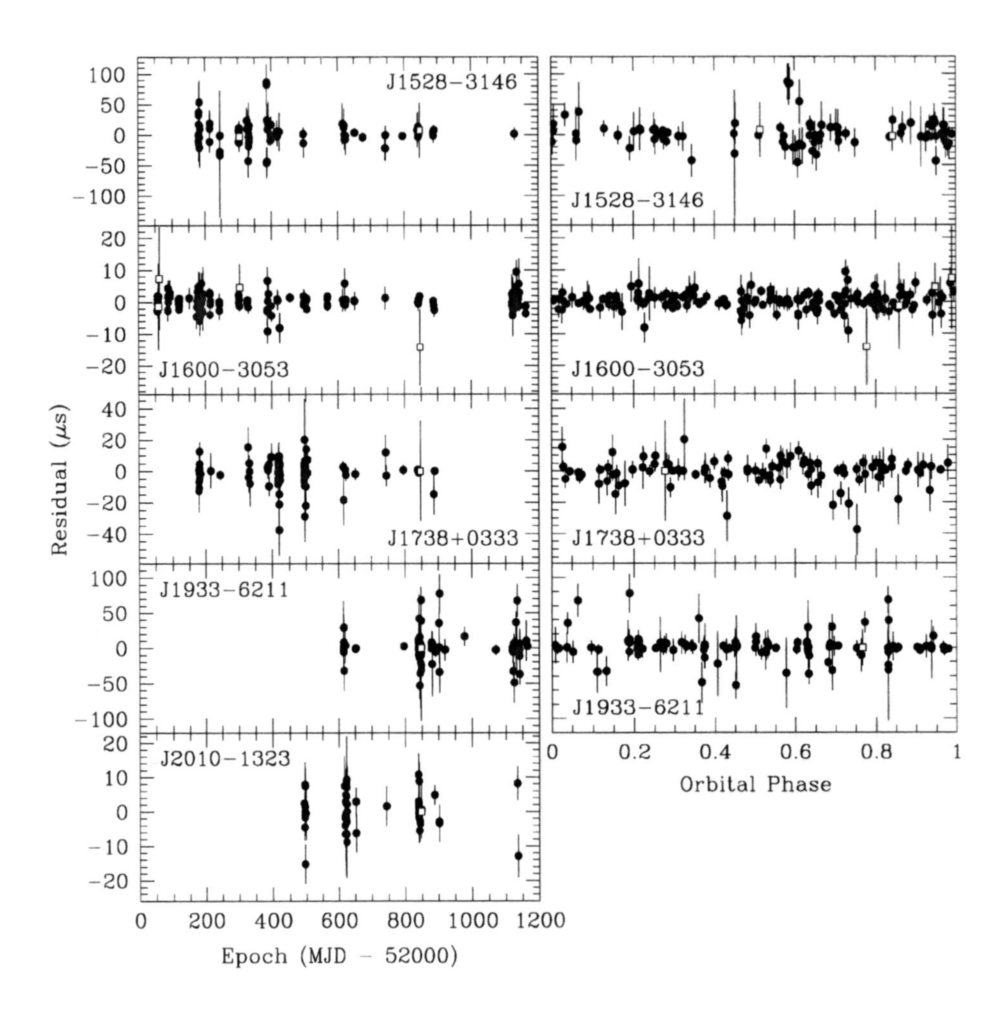

Figure 3.2: Timing residuals plotted versus observation epoch (left column) and orbital phase (right column) for pulsars with phase-connected timing solutions. Filled circles represent observations at 1390 MHz, open squares represent 600 MHz observations.

be suitable for high precision timing experiments, particularly using the 305-m Arecibo telescope.

We found a faint blue object ($V \approx 21$) at the pulsar's position in the digitized sky survey plates; subsequent imaging and spectroscopy show that it is a DA white dwarf with $T_{\rm eff} \approx 8500$ K. On the night of 3 July 2003, we obtained five spectra of this counterpart spanning about half of the system's orbit with the Echellette Spectrograph and Imager (Sheinis et al., 2002) on the 10-m Keck II telescope. The spectra were reduced using standard techniques with the MAKEE[2] package. Radial velocities were obtained by cross correlation of the observed spectra with a template; these radial velocities were corrected to the barycentric frame in both time and velocity. Even with this preliminary effort, the orbital motion is clearly detected (Fig. 3.3). The amplitude of the companion's radial velocity variations is $181 \pm 27\,{\rm km\,s^{-1}}$, and the systemic radial velocity is $30 \pm 21\,{\rm km\,s^{-1}}$. This implies a mass ratio of 8.6 ± 1.6 for the system. Both the radial and transverse velocities of the system are relatively small, giving a 3-D velocity of $42 \pm 26\,{\rm km\,s^{-1}}$, neglecting the uncertainty in the distance to the system.

3.3.4 PSR J1741+13: LMBP with Strong Scintillation

This 3.7 ms LMBP scintillates very strongly at 1.4 GHz, frequently making it undetectable in reasonable integration times at Parkes. In fact, four attempts were required to confirm this pulsar candidate. This strong scintillation likely explains why this pulsar was not discovered in previous Arecibo surveys and suggests that multiple survey observations over a given area of sky may be a worthwhile strategy. We have not yet obtained a phase-connected timing solution for this pulsar.

3.3.5 PSR J1933−6211: LMBP or IMBP?

This short period pulsar's companion has a minimum mass of $0.32\,M_\odot$ — somewhat higher than the typical LMBP. The only other pulsars with companions of at least $0.3\,M_\odot$ and spin periods shorter than 10 ms are PSR J2019+2425 (Nice et al., 1993) with a much longer orbital period, and PSR J1757−5322 (Edwards & Bailes, 2001b) and PSR J1435−6100 (Camilo et al., 2001) with substantially more compact orbits.

---

[2]http://spider.ipac.caltech.edu/staff/tab/makee/

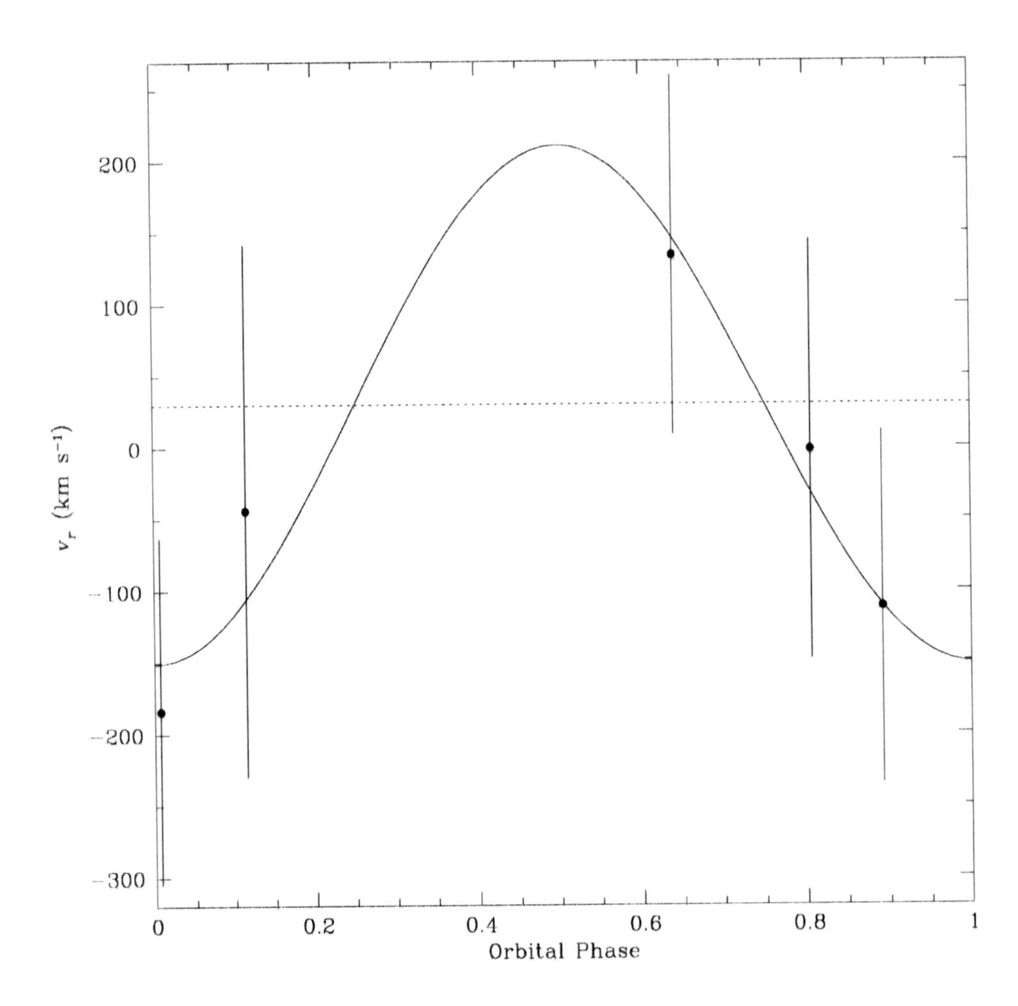

Figure 3.3: Radial velocity of PSR J1738+0333 companion as a function of orbital phase. The curve represents the best-fit model, holding $T_{\rm asc}$ and P_b fixed at the values derived from pulse timing. The dotted line indicates the best-fit systemic radial velocity.

3.3.6 PSR J2010−1323: Isolated MSP

This 5.2 ms pulsar is the only isolated MSP found in this survey. Approximately one-fourth of recycled pulsars not associated with globular clusters are isolated.

We thank R. Edwards for invaluable help with pulsar search software, and D. Fox for taking the Keck observations. BAJ thanks NSF and NASA for supporting his research. The Parkes telescope is part of the Australia Telescope which is funded by the Commonwealth of Australia for operation as a National Facility managed by CSIRO. Some of the data presented herein were obtained at the W.M. Keck Observatory, which is operated as a scientific partnership among the California Institute of Technology, the University of California, and the National Aeronautics and Space Administration. The Observatory was made possible by the generous financial support of the W.M. Keck Foundation.

Chapter 4

PSR J1909−3744: A Binary Millisecond Pulsar with a Very Small Duty Cycle†

B. A. Jacoby[a], M. Bailes[b], M. H. van Kerkwijk[c], S. Ord[b], A. Hotan[b,d], S. R. Kulkarni[a], and S. B. Anderson[a]

[a]Department of Astronomy, California Institute of Technology, MS 105-24, Pasadena, CA 91125; baj@astro.caltech.edu, srk@astro.caltech.edu, sba@astro.caltech.edu.

[b]Centre for Astrophysics and Supercomputing, Swinburne University of Technology, P.O. Box 218, Hawthorn, VIC 31122, Australia; mbailes@swin.edu.au, sord@swin.edu.au, ahotan@swin.edu.au.

[c]Department of Astronomy and Astrophysics, University of Toronto, 60 St. George Street, Toronto, ON M5S 3H8, Canada; mhvk@astro.utoronto.ca.

[d]Australia Telescope National Facility, CSIRO, P.O. Box 76, Epping, NSW 1710, Australia.

Abstract

We report the discovery of PSR J1909−3744, a 2.95 millisecond pulsar in a nearly circular 1.53 day orbit. Its narrow pulse width of 43 μs allows pulse arrival times to be determined with great accuracy. We have spectroscopically identified the companion as a moderately hot ($T \approx 8500$ K) white dwarf with strong absorption lines. Radial velocity measurements of the companion will yield the mass ratio of the system. Our timing data suggest the presence of Shapiro delay; we expect that further timing observations, combined with the mass ratio,

†A version of this chapter was published in *The Astrophysical Journal*, vol. 599, L99–L102, (2003).

will allow the first accurate determination of a millisecond pulsar mass. We have measured the timing parallax and proper motion for this pulsar which indicate a transverse velocity of 140^{+80}_{-40} km s^{-1}. This pulsar's stunningly narrow pulse profile makes it an excellent candidate for precision timing experiments that attempt to detect low frequency gravitational waves from coalescing supermassive black hole binaries.

4.1 Introduction

Binary radio pulsars provide a rich laboratory for a wide range of physical inquiry, including neutron star masses and the evolution of binary systems. The pulsars with the best-determined masses all have eccentric relativistic orbits, spin periods of tens of milliseconds, and masses clustered very tightly around 1.35 M$_\odot$ (Thorsett & Chakrabarty, 1999; Bailes et al., 2003). It is generally thought that millisecond pulsars ($P \leq 10$ ms) should have greater mass than these longer-period pulsars due to a larger accreted mass. There is indeed a statistical suggestion that this is the case, but to date, no precise millisecond pulsar mass measurements exist. Currently, the best-determined millisecond pulsar masses are $m_p = 1.57^{+0.12}_{-0.11}\,M_\odot$ for PSR B1855+09 (Nice et al., 2003) and $m_p = 1.58 \pm 0.18\,M_\odot$ for PSR J0437−4715 (van Straten et al., 2001).

PSR J1909−3744 was discovered during the Swinburne High Latitude Pulsar Survey, a recently-completed pulsar search using the 13-beam multibeam receiver on the 64-m Parkes radio telescope (Jacoby 2003; Jacoby et al., in preparation, see Chapter 2). This survey is similar to, and an extension of, the highly-successful Swinburne Intermediate Latitude Pulsar Survey (Edwards et al., 2001; Edwards & Bailes, 2001a,b). The pulsar was discovered in three adjacent survey beams observed on 2001 January 25 and confirmed on 2001 May 26. Due to the pulsar's strong broadband scintillation, the initial position inferred from the discovery data was in error by nearly a full beamwidth; it was not until we obtained a phase-connected timing solution that we realized the correct position.

Among the properties that can make a pulsar useful for precision timing experiments are a bright, narrow pulse profile, extremely stable rotation, proximity to the earth, and a companion star which can be detected and studied. PSR J1909−3744 appears to possess all of these desirable traits. Nearly all of the pulsar's flux comes in a single narrow, sharp peak with full-width half-maximum (FWHM) of only 43 μs, less than 1.5% of the 2.95 ms pulse period (Fig. 4.1). The brightness of the pulsar varies by at least a factor of 30 in the 21-cm band. The mean flux density of our observations is 3 mJy, but this is biased by the fact that we tend to observe longer when the pulsar is bright. Extremely high-precision timing observations are possible during episodes of favorable scintillation.

4.2 Pulse Timing

Following the confirmation of PSR J1909−3744, we began timing observations with the $2 \times 512 \times 0.5$ MHz Parkes filterbank centered on 1390 MHz. As a result of our incorrect discovery position, few observations taken before February 2002 yielded data useful for high-precision timing. Integration times ranged from three to 100 minutes, with the longer observations broken into 10-minute sub-integrations.

We followed standard pulsar timing procedures: folded pulse profiles from individual observations were cross-correlated with a high signal-to-noise template profile to determine an average pulse time of arrival (TOA) corrected to UTC(NIST). The standard pulsar timing package TEMPO[1], along with the Jet Propulsion Laboratory's DE405 ephemeris, was used for all timing analysis. All 1.4 GHz TOAs with uncertainty less than $1\,\mu$s were included. TOA uncertainties were multiplied by 1.25 to achieve reduced $\chi^2 \simeq 1$. Several observations were also obtained in the 50-cm band with the $2 \times 256 \times 0.125$ MHz Parkes filterbank centered on 660 MHz. As these observations did not yield TOA uncertainties less than $1\,\mu$s, they were used only to determine the dispersion measure (DM) which was then held fixed.

Because of the system's low eccentricity (e), we used the ELL1 binary model which replaces the familiar longitude of periastron (ω), time of periastron (T_0), and e with the time of ascending node ($T_{\rm asc}$), and the Laplace-Lagrange parameters $e \sin \omega$ and $e \cos \omega$ (Lange et al., 2001). We give the results of our timing analysis in Table 7.1 and show residuals relative to this best-fit model in Figure 5.1. The weighted RMS residual is only 330 ns — among the best obtained for any pulsar over ~2 year timespans.

We have begun regular observations of PSR J1909−3744 with the Caltech-Parkes-Swinburne Recorder II (CPSR2; see Appendix B and Bailes, 2003). CPSR2 is a baseband recorder and real-time coherent dedispersion system with two dual-polarization bands, each

[1] http://pulsar.princeton.edu/tempo/

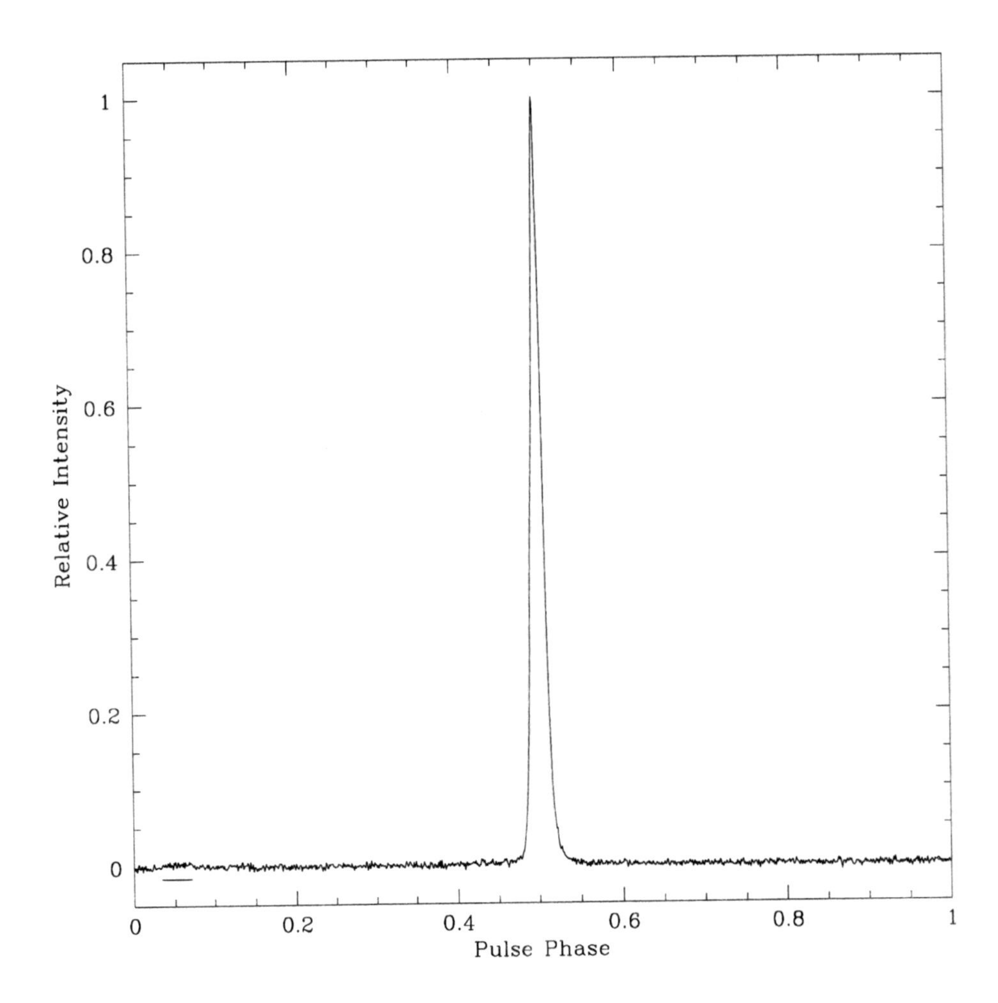

Figure 4.1: Coherently-dedispersed average pulse profile of PSR J1909−3744 at 1341 MHz as measured by the CPSR2 baseband system. Instrumental smearing is just 2μs. The location of a very weak interpulse is indicated by the underscore at left.

Table 4.1. Parameters of the PSR J1909−3744 system

Parameter	Value[a]
Right ascension, α_{J2000}	$19^{h}09^{m}47^{s}.44008(2)$
Declination, δ_{J2000}	$-37^{\circ}44'14''.226(1)$
Proper motion in α, μ_{α} (mas yr^{-1})	−9.6(2)
Proper motion in δ, μ_{δ} (mas yr^{-1})	−35.6(7)
Annual parallax, π (mas)	1.22(44)
Pulse period, P (ms)	2.9471080205034(1)
Period derivative, $\dot{P}$ (10^{-20})	1.4026(3)
Reference epoch (MJD)	52055.8704
Dispersion measure, DM (pc cm^{-3})	10.394(1)
Binary period, P_b (d)	1.5334494503(1)
Projected semimajor axis, $a \sin i$ (lt-s)	1.897992(4)
$e \sin \omega$ ($\times 10^{-7}$)	0.1(90)
$e \cos \omega$ ($\times 10^{-7}$)	−2.6(17)
Time of ascending node, T_{asc} (MJD)	52055.87046667(6)
$\sin i$	1.1(4)
Companion mass m_c ($M_{\odot}$)	0.13(37)
Derived Parameters	
Mass function, $f(m)$	0.00312196(2)
Orbital eccentricity, e ($\times 10^{-7}$)	2.6(18)
Longitude of periastron, ω (deg)	177.874428 ± 190
Time of periastron, T_0	52056.6281373 ± 0.8
Parallax distance, d_{π} (kpc)	$0.82^{+0.46}_{-0.22}$
Transverse velocity, $v_{\perp}$ (km s^{-1})	140^{+80}_{-40}
Intrinsic period derivative, $\dot{P}_{int}$ (10^{-20})[b]	$0.61^{+0.23}_{-0.50}$
Surface magnetic field, B_{surf} ($\times 10^{8}$G)[b]	$1.3^{+0.2}_{-0.8}$
Characteristic age, τ_c (Gyr)[b]	$7.7^{+35.0}_{-2.1}$
Galactic longitude, l (deg)	359.73
Galactic latitude, b (deg)	−19.60
Distance from Galactic plane, $\|z\|$ (kpc)	$0.28^{+0.16}_{-0.07}$
Pulse FWHM, w_{50} (μs)	43
Pulse width at 10% peak, w_{10} (μs)	89

[a]Figures in parenthesis are uncertainties in the last digit quoted. Uncertainties are calculated from twice the formal error produced by TEMPO.

[b]Corrected for secular acceleration based on measured proper motion and parallax

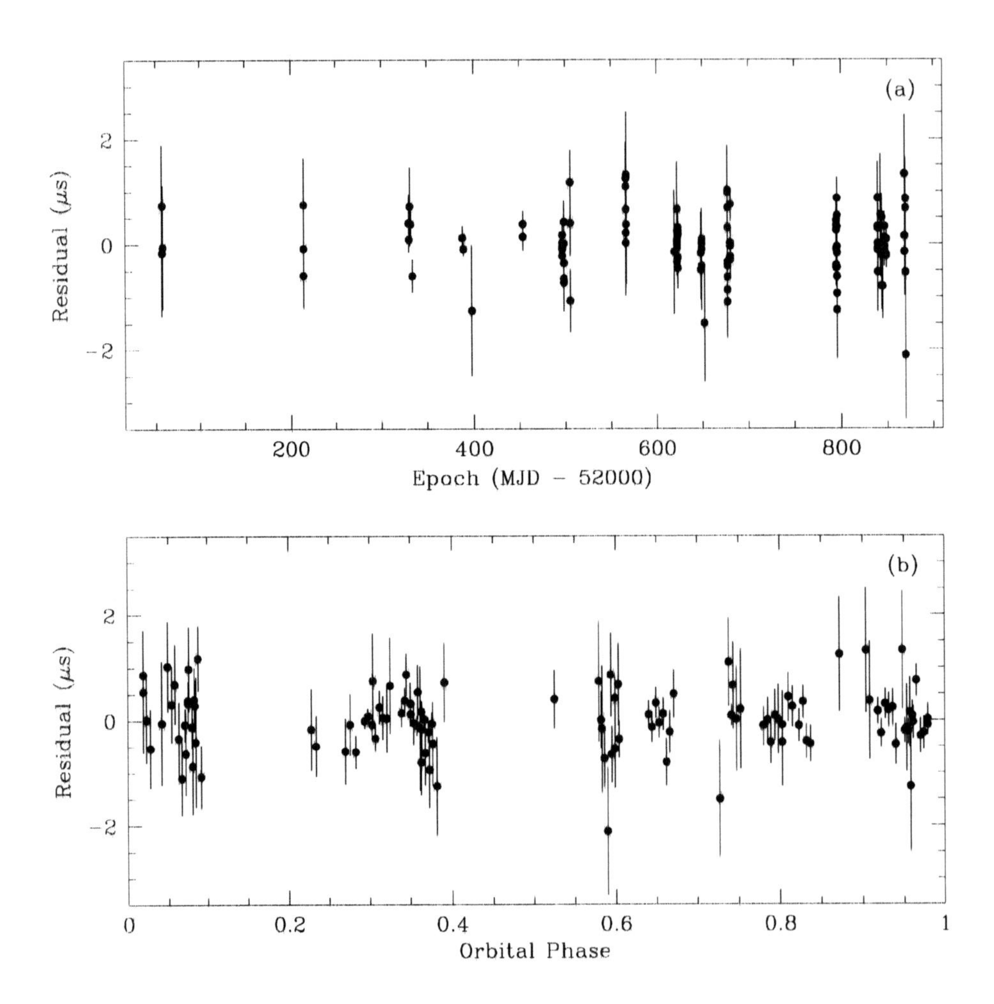

Figure 4.2: Timing residuals for PSR J1909−3744. (a): Residuals vs. observation epoch. (b): Residuals vs. orbital phase. Inferior conjunction occurs at orbital phase 0.16.

64 MHz wide. The CPSR2 pulse profile shown in Figure 4.1 is smeared by only 2 μs, compared with the 82 μs time resolution of the 512-channel filterbank. Initial results from CPSR2 data show great promise for high precision timing, and will be reported in a future paper (Chapter 5).

4.2.1 Shapiro Delay

Though our current data set suffers from poor orbital phase coverage (specifically, a lack of high-quality observations near inferior conjunction), we find strong evidence of Shapiro delay. If we remove Shapiro delay (i.e., companion mass, m_c, and sine of orbital inclination, $\sin i$) from the timing model and allow TEMPO to optimize the remaining parameters, χ^2 increases from 98.8 to 128.1 (for 102 and 104 degrees of freedom, respectively), indicating a highly significant detection. The peak-to-peak residual is essentially the same in either case (3.4 μs) as much of the Shapiro delay signal is absorbed into an erroneously large Roemer delay; however, the weighted RMS residual increases to 370 ns when the model excludes m_c and $\sin i$. Holding the astrometric, orbital, and spin parameters fixed at the values in Table 7.1 but neglecting Shapiro delay increases the peak-to-peak residual to 8.1 μs, with pulses observed nearest inferior conjunction arriving later than predicted by the model.

Given its large uncertainty, we are not concerned by the most likely value of $\sin i > 1$ found by TEMPO's linear least-squares fit. To confirm that the Shapiro delay is physically meaningful, we constructed a χ^2 map as a function of m_c and $\sin i$, while optimizing the remaining parameters. Our sparse data set is consistent with a large range of allowed parameter space, constraining the system to have $\sin i \geq 0.85$ and $0.17\,M_\odot \leq m_c \leq 0.55\,M_\odot$ (95% confidence). The covariance of the poorly-determined m_c and $\sin i$ with the projected semimajor axis ($a \sin i$) and e means that the uncertainties in these parameters given in Table 7.1 are likely to be underestimated.

Currently, the strongest constraint on the companion mass is the minimum mass of $0.195\,\mathrm{M}_\odot$ obtained from the mass function by assuming $i = 90°$ and $m_p = 1.4\,\mathrm{M}_\odot$; the fact that Shapiro delay is detected at all suggests that we are likely viewing the system roughly edge-on. Future pulsar timing observations near inferior conjunction will significantly improve the Shapiro delay-derived companion mass.

4.2.2 Distance

We have measured the parallax, $\pi = 1.22 \pm 0.44$ mas, and hence can estimate the distance to PSR J1909−3744, $d_\pi = 0.82^{+0.46}_{-0.22}$ kpc. Combined with the proper motion, this distance corresponds to a transverse velocity of $140^{+80}_{-40}\,\mathrm{km\,s^{-1}}$. The DM-derived distance estimate is 0.46 kpc (Cordes & Lazio, 2002).

The pulsar's proper motion induces an apparent radial acceleration equal to $\mu^2 d/c$ (Shklovskii, 1970). This secular acceleration will allow an entirely independent distance estimate from precise measurement of the orbital period derivative, $\dot{P}_b$ (Bell & Bailes, 1996). For circular binaries such as PSR J1909−3744, the intrinsic $\dot{P}_b$ due to general relativity is negligible. However, the secular acceleration will eventually produce a measurable $\dot{P}_b$ which can be used to determine the distance. The peak-to-peak timing signal due to parallax is $\sim 1.4\,\mu$s, while the $\dot{P}_b$ signal increases with the observing baseline, t, as $\sim 240\,\mathrm{ns}\,(t/1\,\mathrm{yr})^2$. Thus, after only a few years of timing $\dot{P}_b$ will give the more precise distance. We note that the acceleration induced by differential Galactic rotation is about two orders of magnitude smaller than the secular acceleration.

The secular acceleration also corrupts the observed period derivative ($\dot{P}$). Since the pulsar's intrinsic spindown rate ($\dot{P}_{\mathrm{int}}$) is non-negative, we can obtain an upper distance limit of 1.4 kpc, consistent with the measured parallax. Conversely, the measured proper motion and parallax-derived distance allow us to correct $\dot{P}$ for the secular acceleration and hence, determine $\dot{P}_{\mathrm{int}}$.

4.3 Optical Observations of the White Dwarf Companion

At the position of the pulsar, the digitized sky survey plates showed a possible counterpart. To verify this, we obtained images on the night of 2003 June 4 with the Magellan Instant Camera (MagIC) at the 6.5-m Clay telescope on Las Campanas. We took a 10-min exposure in B and a 5-min exposure in R. The conditions were poor, with thin clouds and $1''$ seeing.

The images were reduced using the Munich image data analysis system (MIDAS), following the usual steps of bias subtraction (separately for the four amplifiers) and flat-fielding using dome flats. For the astrometric calibration, we selected 58 USNO-B1.0 (Monet et al., 2003) objects that overlapped with our images, were not overexposed, and appeared stellar. For these objects, we measured centroids and fitted for zero-point position, plate scale, and

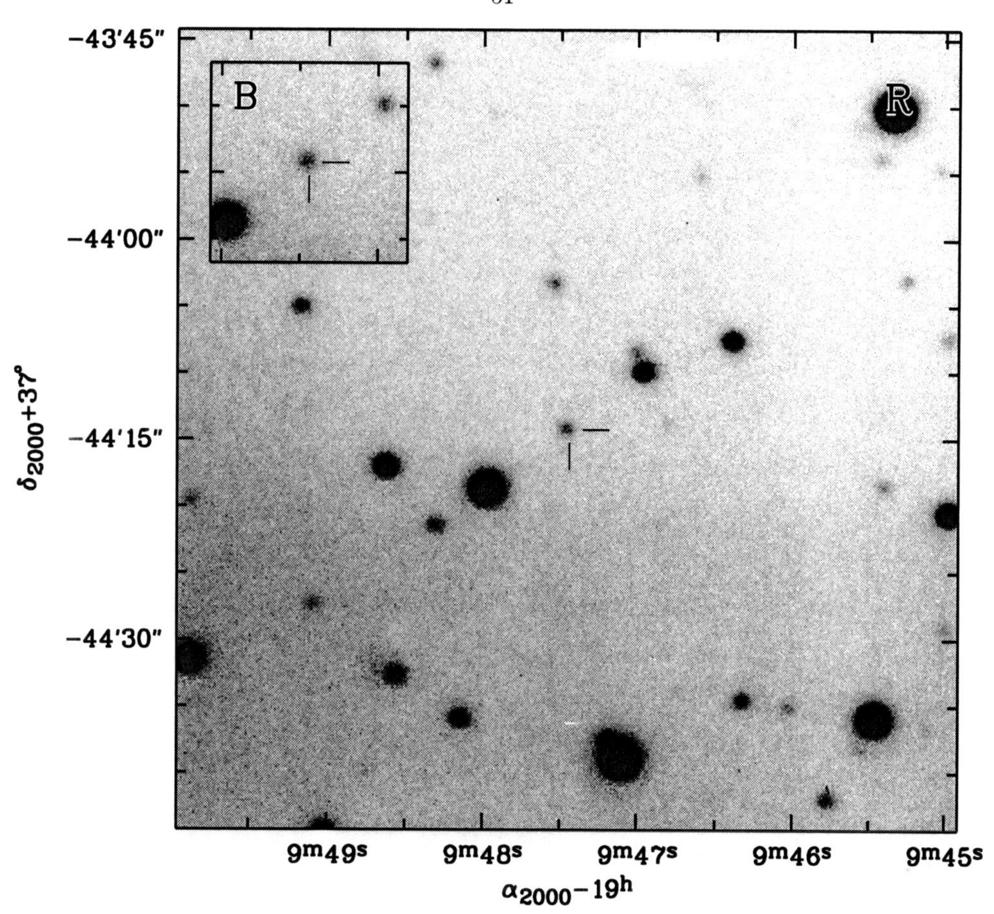

Figure 4.3: Image of the field of PSR J1909−3744. The large frame shows our R-band image; the position of the pulsar is indicated by the tick marks. The inset shows the B-band image; compared to other stars in the field, the companion is blue.

position angle. Rejecting 23 outliers with residuals larger than $0\farcs5$, the inferred single-star measurement error in both bands is $0\farcs15$ in each coordinate, which is consistent with expectations for the USNO-B1.0 measurements. Hence our astrometry should be tied to the USNO-B1.0 system at the $\sim 0\farcs03$ level.

In both images, we found an object near the position of the pulsar which is blue relative to other stars in the field (see Fig. 4.3). The position inferred from our astrometry is $\alpha_{\rm J2000} = 19^{\rm h}09^{\rm m}47\fs457$, $\delta_{\rm J2000} = -37^\circ44'14\farcs17$, which is consistent with the timing position listed in Table 7.1 within the uncertainty with which the USNO-B1.0 system is tied to the International Celestial Reference Frame ($\sim 0\farcs2$ in each coordinate).

The identification of this object with the pulsar's companion was confirmed by spectra, taken on the night of 2003 June 6, using the Low Dispersion Survey Spectrograph 2 (LDSS2) at the Clay telescope. We took two 30-min exposures with the high resolution grism and a 1″ slit, which gives a resolution of ~ 6 Å. The conditions were mediocre, with thin cirrus and 0.″7 seeing.

We used MIDAS to reduce the spectra. They were bias-subtracted and flat-fielded using normalized dome flats (in which the noisy blue part was set to unity). After sky subtraction, the two spectra were extracted using optimal weighting and added together. To derive the response, we used spectra of Feige 110 taken during the same night, and fluxes from calibrated STIS spectra.[2] The result is shown in Fig. 4.4. We stress that since cirrus was present, the absolute flux calibration is not reliable, although both from the spectrum and from a rough photometric calibration using USNO-B1.0 magnitudes, it follows that the companion has $V \simeq 21$.

The spectrum (Fig. 4.4) shows only the Balmer lines of Hydrogen, from Hα up to H10, consistent with a DA white dwarf of low surface gravity (and thus low mass). The strength of the lines and the slope of the spectrum are similar to what is seen for PSR J1012+5307 (van Kerkwijk et al., 1996) and PSR J0218+4232 (Bassa et al., 2003), indicating a similar temperature of ~ 8500 K.[3]

Assuming a mass of $\sim 0.20\,M_{\odot}$, the white dwarf should have a radius of approximately $0.03\,R_{\odot}$ and a luminosity of $\sim 4 \times 10^{-3}\,L_{\odot}$ (Driebe et al., 1998; Rohrmann et al., 2002). The corresponding absolute magnitude is $M_V \simeq 11.0$, which implies $V \simeq 20.5$ at a distance of 0.8 kpc, consistent with the rough photometry quoted above.

4.4 Conclusions

Pulse timing precision is often limited by systematic errors that distort the shape of pulsar profiles due to the interstellar medium and imperfect polarimetric and instrumental response. Intrinsically narrow pulses are less susceptible to these errors, as are pulsars with small dispersion measures. PSR J1909−3744 is likely to become a cornerstone of future timing array experiments to detect low-frequency gravitational waves. It is noteworthy that

[2] See http://www.stsci.edu/instruments/observatory/cdbs/calspec.html.

[3] This high temperature may be surprising, given the pulsar's long characteristic age of roughly 8 Gyr. It indeed sets interesting constraints on the thickness of the outer Hydrogen envelope of the white dwarf; for a discussion and references, see Bassa et al. (2003).

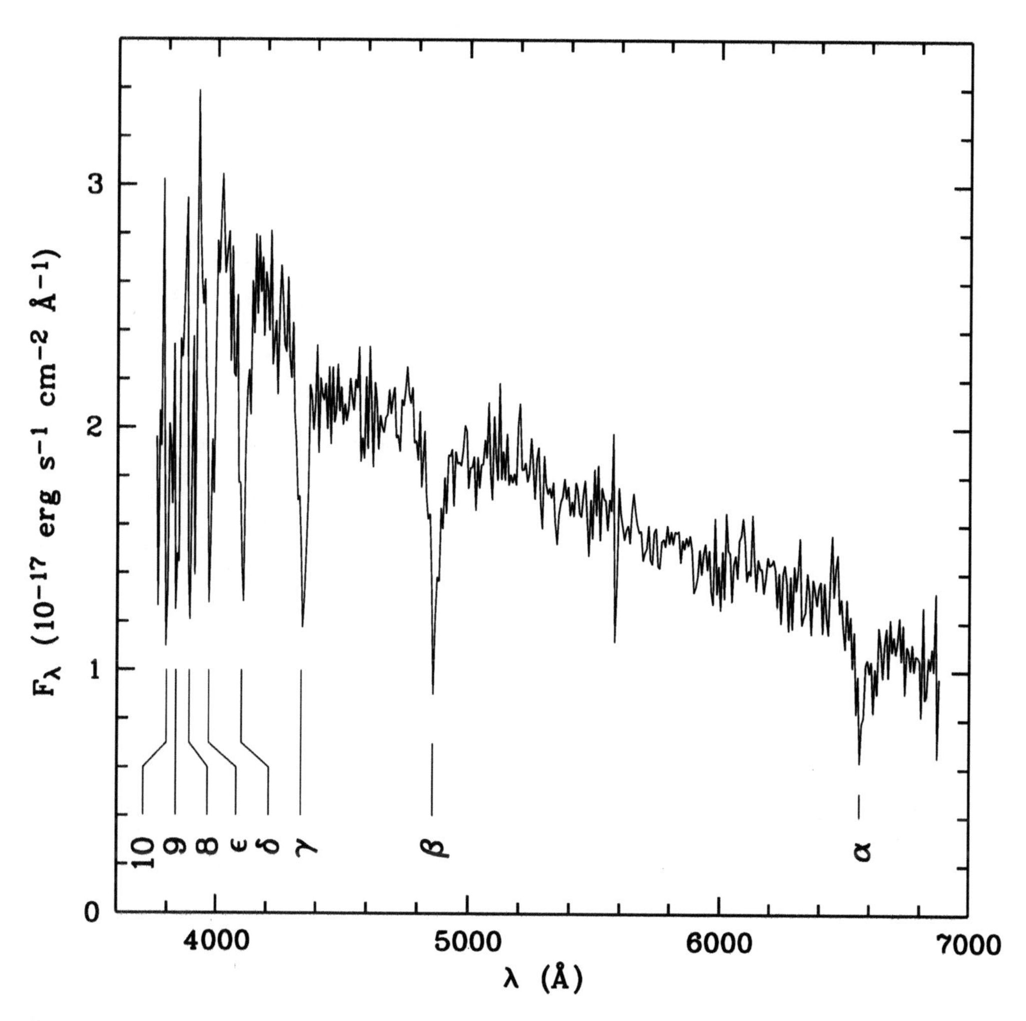

Figure 4.4: Spectrum of the companion of PSR J1909−3744, showing that it is a low-mass DA white dwarf. The Balmer lines from Hα to H10 are indicated. Note that the absolute flux calibration is uncertain by about 50%, as cirrus was present during the observations. The relative calibration, however, should be reliable (except longward of ~ 6400 Å, where the spectrum might be contaminated by second-order light).

PSR J1909−3744 is far from other high-precision pulsars on the sky and therefore probes a unique line of sight.

In the shorter run, further timing of the pulsar, especially with CPSR2, should yield an accurate inclination and a reasonably precise companion mass. Furthermore, in a modest spectroscopic observing campaign on an 8-m class telescope, the radial velocity orbit of the companion can be measured precisely. These combined results should allow a very reliable pulsar mass measurement and advance our understanding of binary pulsar evolution and recycling.

We thank R. Edwards for invaluable help with pulsar search software, H. Knight for assistance with Parkes observations, and M. Hamuy, V. Mariño, and M. Roth for help with the Magellan observations. BAJ and SRK thank NSF and NASA for supporting their research. MHvK acknowledges support by the National Sciences and Engineering Research Council of Canada. The Parkes telescope is part of the Australia Telescope which is funded by the Commonwealth of Australia for operation as a National Facility managed by CSIRO.

Chapter 5

The Mass of a Millisecond Pulsar†

B. A. Jacoby[a], A. Hotan[b,c], M. Bailes[b], S. Ord[b], and S. R. Kulkarni[a]

[a]Department of Astronomy, California Institute of Technology, MS 105-24, Pasadena, CA 91125; baj@astro.caltech.edu, srk@astro.caltech.edu

[b]Centre for Astrophysics and Supercomputing, Swinburne University of Technology, P.O. Box 218, Hawthorn, VIC 31122, Australia; ahotan@swin.edu.au, mbailes@swin.edu.au, sord@swin.edu.au.

[c]Australia Telescope National Facility, CSIRO, P.O. Box 76, Epping, NSW 1710, Australia.

Abstract

We report on nearly two years of high-precision timing observations of the low-mass binary millisecond pulsar, PSR J1909−3744. The edge-on orientation of the system's orbit, combined with the pulsar's small duty cycle and extremely stable rotation, have allowed measurement of Shapiro delay and the most precise determination to date of a millisecond pulsar mass, $m_p = (1.438 \pm 0.024)\, M_\odot$. All previous neutron star mass measurements of similar or better precision have been in eccentric systems with massive companions; these pulsars have spin periods roughly an order of magnitude longer than PSR J1909−3744 and are expected to have accreted less material from their companions in the recycling process. The mass of PSR J1909−3744 is at the upper edge of the mass range observed in double neutron star binaries. We have greatly improved upon the previous parallax and proper motion measurements of PSR J1909−3744, yielding a distance of $(1.14^{+0.08}_{-0.07})$ kpc and transverse velocity of $(200^{+14}_{-13})\,\mathrm{km\,s^{-1}}$. Our timing measurements of PSR J1909−3744 are the most precise ever obtained over ~2 year time spans.

†Part of a manuscript in preparation for publication.

5.1 Introduction

Radio pulsars in binary systems can allow for the measurement of neutron star masses, thereby providing one side of the dense matter equation of state. However, to date all precisely-measured pulsar masses have been relatively slowly rotating objects (spin period P typically tens of milliseconds) in double neutron star systems or in one case, an eccentric pulsar-massive white dwarf system (see Stairs, 2004 for a recent review). These neutron stars all fall within the narrow mass range of $(1.35 \pm 0.04)\,M_\odot$ established by Thorsett & Chakrabarty (1999), with the exception of PSR B1913+16 at $(1.4408 \pm 0.0003)\,M_\odot$ (Weisberg & Taylor, 2003) and PSR J0737−3039B at $(1.250 \pm 0.005)\,M_\odot$ (Lyne et al., 2004).

Millisecond pulsars ($P < 10\,\mathrm{ms}$) are generally thought to have accreted more matter from a low-mass binary companion and therefore, are expected to be more massive than the slower-rotating recycled pulsars with massive companions. However, these low-mass binary pulsars do not experience measurable relativistic orbital period decay ($\dot{P}_b$) and their typically circular orbits make orbital precession ($\dot{\omega}$) and time dilation and gravitational redshift (γ) difficult to measure; therefore, mass determinations have been less precise. There is a statistical suggestion that millisecond pulsars are indeed more massive than slower-rotating neutron stars (Kaspi et al., 1994; van Straten et al., 2001; Freire et al., 2003), but with the exception of PSR J0751+1807 at $(2.1 \pm 0.2)\,M_\odot$ (Nice et al. in preparation; Nice et al., 2004) and possibly PSR J1748-2446I in the globular cluster Terzan 5 (Ransom et al., in preparation), all are still consistent with the observed mass distribution of pulsars with massive companions.

PSR J1909−3744 was discovered during a large-area survey for pulsars at high galactic latitudes with the 64-m Parkes radio telescope (Jacoby et al., 2003, see Chapters 2 and 4). It is a typical low-mass binary pulsar with $P = 2.95\,\mathrm{ms}$, but with several exceptional qualities: its extremely narrow pulse profile (Ord et al., 2004) and stable rotation allow us to time it with precision rivaled by few other pulsars, and its nearly edge-on orbital inclination means the pulsar signal experiences strong Shapiro delay. Measurement of this Shapiro delay gives the orbital inclination and companion mass to high precision, and combined with the Keplerian orbital parameters, allow us to measure the pulsar's mass.

5.2 Observations and Pulse Timing

In December 2002, we began observing PSR J1909−3744 with the Caltech-Parkes-Swinburne Recorder II (CPSR2, see Appendix B and Bailes, 2003) at the Parkes radio telescope. This instrument Nyquist samples the voltage signal from the telescope in each of two 64-MHz wide dual-polarization bands with 2-bit precision. For these observations, we used the center beam of the Parkes Multibeam receiver or H-OH receiver and placed the two bands at sky frequencies of 1341 MHz and 1405 MHz. The raw voltage data were sent to a dedicated cluster of 30 dual-processor Pentium IV computers for immediate analysis. The two-bit sampled data were corrected for quantization effects (Jenet & Anderson, 1998) and coherently dedispersed into 128 frequency channels in each of 4 Stokes parameters which were then folded at the topocentric pulse period using the PSRDISP software package (van Straten, 2002). We recently began observing PSR J1909−3744 with the Caltech-Green Bank-Swinburne Recorder II (CGSR2), a clone of CPSR2 installed at the 100-m Green Bank Telescope (GBT). As the GBT data set covers only a small fraction of the system's orbital phase, we have not included it in this analysis.

Off-line data reduction and calculation of average pulse times of arrival (TOAs) were accomplished in the usual manner using the PSRCHIVE[1] suite. Because of roll-off of the anti-aliasing filters, 8 MHz was removed from each band edge prior to formation of dedispersed total intensity profiles, giving a final bandwidth of 48 MHz per band. To avoid averaging over phenomena which vary on orbital timescales, observations longer than 10 minutes were broken into 10-minute segments. Finally, TOAs were calculated by cross-correlation with a high signal-to-noise template profile, formed by summing a total of 5.4 days of integration in the 1341 MHz band. Arrival times with uncertainty greater than 1 μs were excluded from further analysis. Our final data set contains 1730 TOAs — roughly half of which come from each of our two frequency bands.

We used the standard pulsar timing package TEMPO[2], along with the Jet Propulsion Laboratory's DE405 ephemeris for all timing analysis. TOAs were corrected to UTC(NIST). Using the TOA uncertainties estimated from the cross-correlation procedure, our best-fit timing model had reduced $\chi^2 \simeq 1.2$, indicating that our arrival time measurements are relatively free of systematic errors. In our final analysis, these TOA uncertainties were

[1] http://astronomy.swin.edu.au/pulsar/software/libraries/

[2] http://pulsar.princeton.edu/tempo/

multiplied by 1.1 to achieve reduced $\chi^2 \simeq 1$ and improve our estimation of uncertainties in model parameters. We give the results of our timing analysis in Table 7.1. The weighted rms residual resulting from this analysis of 10-minute integrations is only 230 ns. By averaging all data from each day with more than 1 hour of observation we can obtain a weighted rms of 90 ns, to our knowledge the most precise long-term timing ever obtained; however, shorter integrations allow for better determination of orbital parameters. We note that PSR J1909−3744 has the smallest orbital eccentricity of any known astronomical orbit.

5.2.1 Shapiro Delay and Component Masses

As shown in Figure 5.1, our timing data display the unmistakable signature of Shapiro delay. Measurement of this relativistic effect has allowed the precise determination of orbital inclination, $i = (86.58^{+0.11}_{-0.10})^\circ$, and companion mass, $m_c = (0.2038 \pm 0.0022)\,M_\odot$. These values were derived from a χ^2 map in $m_c - \cos i$ space (Fig. 5.2), but are in excellent agreement with the results of TEMPO's linear least-squares fit for m_c and $\sin i$.

Combined with the mass function, our tight constraints on m_c and i determine the pulsar mass, $m_p = (1.438 \pm 0.024)\,M_\odot$ (Fig. 5.3). This result is several times more precise than the previous best mass measurements of heavily recycled neutron stars.

We have used Shapiro delay to measure two post-Keplerian parameters for this sytem, shape $s \equiv \sin i = 0.99822 \pm 0.00011$ and range $r \equiv m_c G/c^3 = (1.004 \pm 0.011)\,\mu\text{s}$, where G is the gravitational constant and c is the speed of light. The value of $\dot{\omega}$ predicted by general relativity is only 0.14 deg yr^{-1}, so it will be many years before a third post-Keplerian parameter can be measured in this extremely circular system.

5.2.2 Distance and Kinematic Effects

The measured parallax, $\pi = (0.88 \pm 0.05)$ mas gives a distance of $d_\pi = (1.14^{+0.08}_{-0.07})$ kpc. We can now calculate the mean free electron density along the line of sight to PSR J1909−3744 based on its dispersion measure (DM), $\langle n_e \rangle \equiv DM/d = DM\pi = (0.0091 \pm 0.0006)\,\text{cm}^{-3}$. The DM-derived distance estimate is 0.46 kpc (Cordes & Lazio, 2002), indicating that the free electron density along the line of sight is significantly overestimated by the model. There are no pulsars near PSR J1909−3744 on the sky with accurate parallax measurements, either through pulse timing or VLBI; PSR J1909−3744 will therefore provide an important constraint on galactic electron density models.

Table 5.1. Improved parameters of the PSR J1909−3744 system

Parameter	Value[a]
Right ascension, α_{J2000}	$19^h09^m47^s.437999(1)$
Declination, δ_{J2000}	$-37°44'14''.31841(8)$
Proper motion in α, μ_α (mas yr^{-1})	−9.47(2)
Proper motion in δ, μ_δ (mas yr^{-1})	−35.8(2)
Annual parallax, π (mas)	0.88(5)
Pulse period, P (ms)	2.947108021647488(6)
Period derivative, $\dot{P}$ (10^{-20})	1.40258(9)
Reference epoch (MJD)	53000.0
Dispersion measure, DM (pc cm^{-3})	10.3939(1)
Binary period, P_b (d)	1.53344945044(2)
Projected semimajor axis, $a \sin i$ (lt-s)	1.89799117(7)
$e \sin \omega$ ($\times 10^{-7}$)	0.56(36)
$e \cos \omega$ ($\times 10^{-7}$)	−1.2(2)
Time of ascending node, T_{asc} (MJD)	53000.475328090(2)
$\sin i$	0.99822(11)
Companion mass m_c ($M_\odot$)	0.2038(22)
Derived Parameters	
Pulsar mass m_p ($M_\odot$)	1.438(24)
Mass function, $f(m)$	0.0031219531(4)
Range, r (μs)	1.004(11)
Orbital inclination, i (deg)	$86.58^{+0.11}_{-0.10}$
Orbital eccentricity, e ($\times 10^{-7}$)	1.4(2)
Longitude of periastron, ω (deg)	155.7452858095 ± 14
Time of periastron, T_0	53001.13873788 ± 0.06
Parallax distance, d_π (kpc)	$1.14^{+0.08}_{-0.07}$
Transverse velocity, $v_\perp$ (km s^{-1})	200^{+14}_{-13}
Intrinsic period derivative, $\dot{P}_{int}$ (10^{-20})[b]	$0.28^{+0.07}_{-0.08}$
Surface magnetic field, B_{surf} ($\times 10^8$G)[b]	$0.92^{+0.11}_{-0.15}$
Characteristic age, τ_c (Gyr)[b]	16^{+7}_{-3}
Galactic longitude, l (deg)	359.73
Galactic latitude, b (deg)	−19.60
Distance from Galactic plane, $\|z\|$ (kpc)	$0.38^{+0.03}_{-0.02}$

[a]Figures in parenthesis are uncertainties in the last digit quoted. Uncertainties are calculated from twice the formal error produced by TEMPO, except as described in §5.2.1.

[b]Corrected for secular acceleration based on measured proper motion and parallax

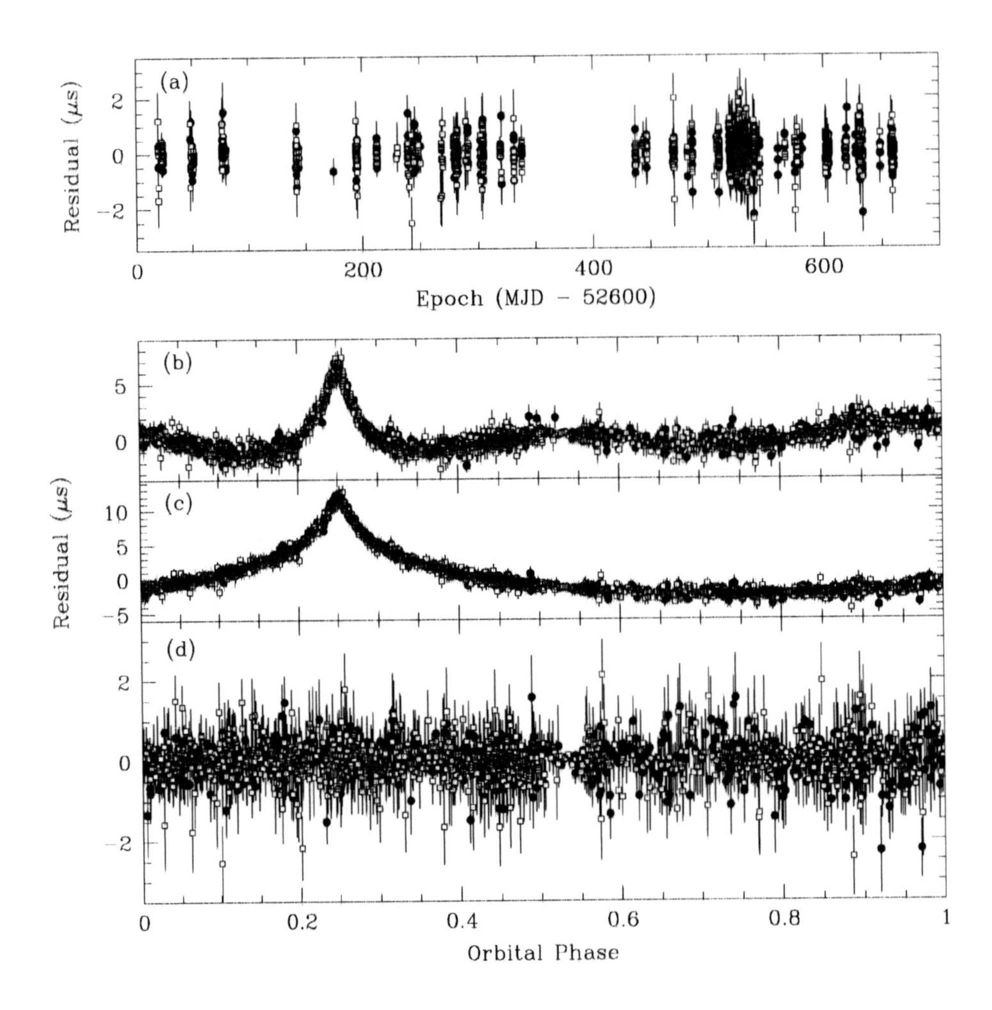

Figure 5.1: High-precision timing residuals for PSR J1909−3744. Filled circles represent TOAs from 1341 MHz band, while open squares denote the 1405 MHz band. (a): Residuals vs. observation epoch for best-fit model taking Shapiro delay fully into account (Tab. 7.1). (b): Residuals vs. orbital phase for best-fit Keplerian model. Some of the Shapiro delay signal is absorbed in an anomalously large Roemer delay and eccentricity. (c): Residuals vs. orbital phase for physically correct Keplerian orbital model, but neglecting Shapiro delay. (d): Residuals vs. orbital phase for the best-fit model taking Shapiro delay fully into account (Tab. 7.1).

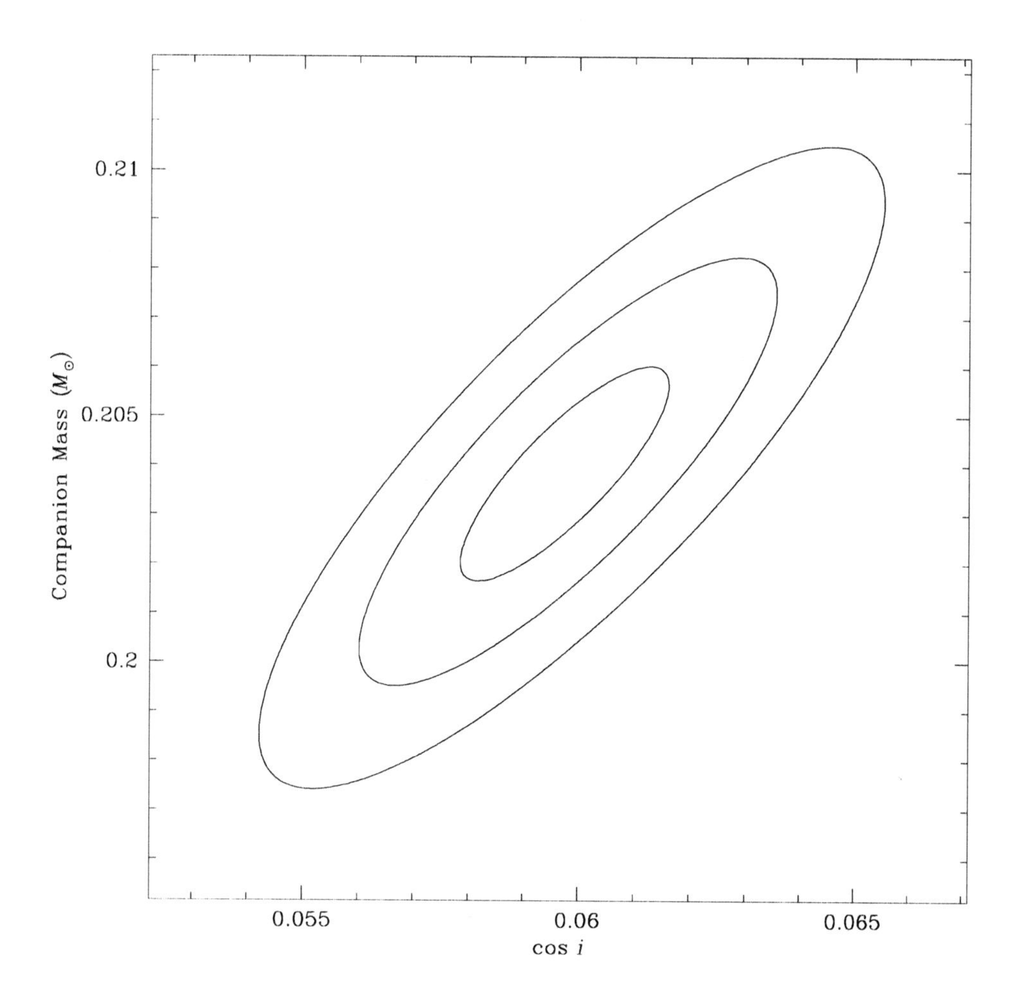

Figure 5.2: Companion mass – orbital inclination diagram for PSR J1909−3744. Contours show $\Delta\chi^2 = 1$, 4, and 9 ($1\,\sigma$, $2\,\sigma$, and $3\,\sigma$) regions, respectively in companion mass and $\cos i$.

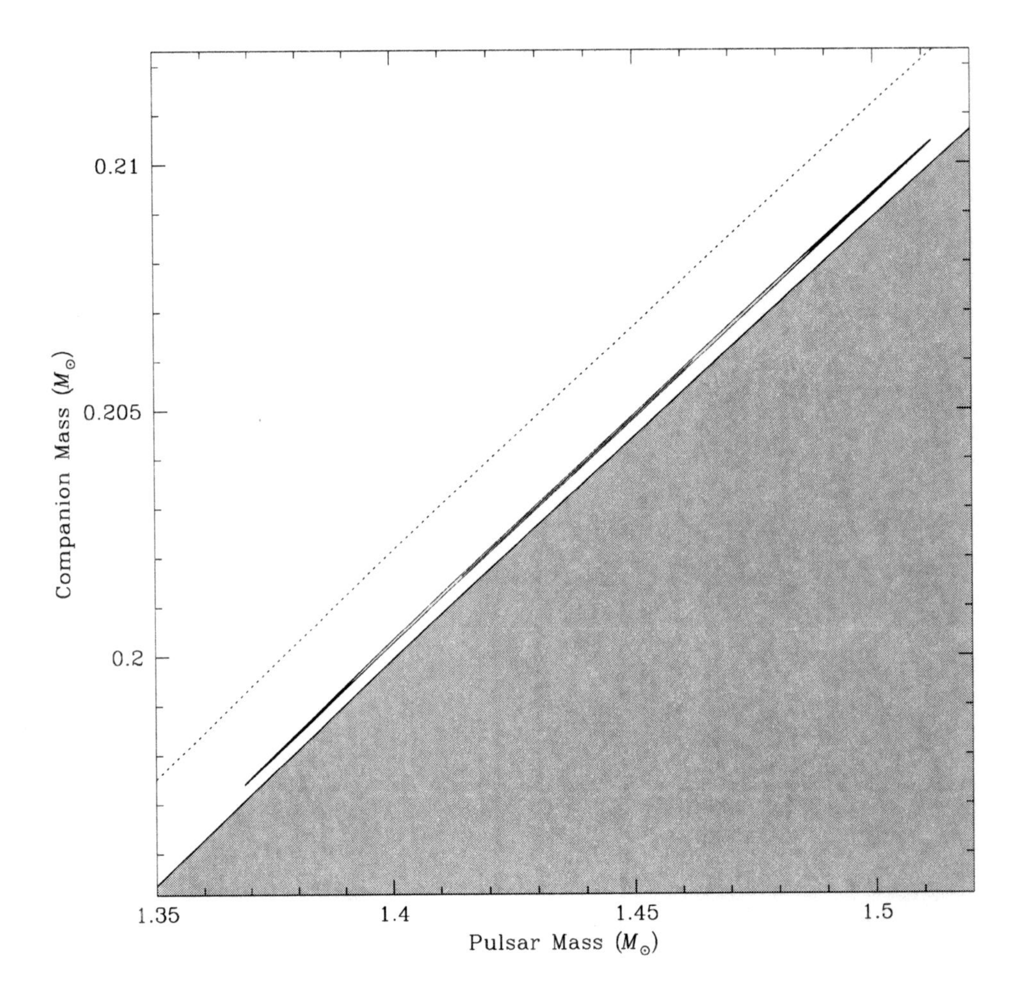

Figure 5.3: Mass – mass diagram for PSR J1909−3744. Red, yellow, and black areas show $\Delta\chi^2 = 1$, 4, and 9 ($1\,\sigma$, $2\,\sigma$, and $3\,\sigma$) regions, respectively in companion mass and pulsar mass. The grey shaded region is excluded by the mass function and the requirement that $\sin i \leq 1$; the dotted line indicates $\sin i = 0.99$ for reference.

The pulsar's proper motion induces an apparent secular acceleration equal to $\mu^2 d/c$ (Shklovskii, 1970). The secular acceleration corrupts the observed period derivative ($\dot{P}$). Since the pulsar's intrinsic spindown rate must be non-negative, we can obtain an upper distance limit of $d_{\rm max} = 1.4\,{\rm kpc}$, consistent with the measured parallax. Conversely, the measured proper motion and parallax-derived distance allow us to correct $\dot{P}$ for the secular acceleration and hence, determine the intrinsic spindown rate (Tab. 7.1). We note that the acceleration induced by differential Galactic rotation is about two orders of magnitude smaller than the secular acceleration and therefore, has been neglected.

Similarly, the secular acceleration induces an apparent $\dot{P}_b$. Based on the distance and proper motion of PSR J1909−3744, this kinematic $\dot{P}_b$ is expected to be $\sim 0.5 \times 10^{-12}$ — about 400 times larger (and of opposite sign) than the predicted intrinsic value. We note that if we include $\dot{P}_b$ in our timing model, the best-fit value is consistent with this prediction but the significance is low. Therefore, we have not included $\dot{P}_b$ in our analysis. However, in several years, this kinematic $\dot{P}_b$ will give an improved measurement of the pulsar distance (Bell & Bailes, 1996).

The transverse velocity resulting from the parallax distance and the measured proper motion is $(200^{+14}_{-13})\,{\rm km\,s^{-1}}$, somewhat higher than typical for binary millisecond pulsars (Toscano et al., 1999).

5.3 Conclusions

We have obtained the most precise measurement of a heavily recycled neutron star through high-precision timing measurements of PSR J1909−3744: $m_p = (1.438 \pm 0.024)\,M_\odot$. While PSR J1909−3744 appears to be more massive than the canonical range for mildly recycled pulsars ($1.35 \pm 0.04\,M_\odot$), its mass is consistent with that of the original member of this class, namely PSR B1913+16. Therefore, it is unlikely that a clear mass distinction can be made between the millisecond pulsars with low-mass white dwarf companions and the mildly recycled pulsars with massive companions.

It is possible that neutron star birth masses are simply not as homogeneous as the sample of measurements suggested until recently. The discovery of PSR J0737−3039B in the double pulsar system with $m_p = 1.25\,M_\odot$ is difficult to reconcile with the millisecond pulsars, PSR J0751+1807 and PSR J1748-2446I, with most probable masses near or exceeding

$2\,M_{\odot}$, based purely on post-supernova accretion history. If we assume that the mass of PSR J0737−3039B (the only well-measured slow pulsar mass and the lowest measured neutron star mass) is representative of pre-accretion neutron stars, it appears that less than $0.2\,M_{\odot}$ is accreted in the process of spinning pulsars up to millisecond periods, implying that most of the companion's original mass is lost from the system.

We acknowledge S. Anderson, W. van Straten, J. Yamasaki, and J. Maciejewski for major contributions to the development of CPSR2, and thank H. Knight for assistance with observations. BAJ and SRK thank NSF and NASA for supporting their research. The Parkes telescope is part of the Australia Telescope which is funded by the Commonwealth of Australia for operation as a National Facility managed by CSIRO.

Chapter 6

Optical Detection of Two Intermediate Mass Binary Pulsar Companions†

B. A. JACOBY[a], D. CHAKRABARTY[b], M. H. VAN KERKWIJK[c], S. R. KULKARNI[a], AND D. L. KAPLAN[a]

[a]Department of Astronomy, California Institute of Technology, MS 105-24, Pasadena, CA 91125; baj@astro.caltech.edu, srk@astro.caltech.edu.

[b]Department of Physics and Center for Space Research, Massachusetts Institute of Technology, Cambridge, MA 02139; deepto@space.mit.edu.

[c]Department of Astronomy and Astrophysics, University of Toronto, 60 St. George Street, Toronto, ON M5S 3H8, Canada; mhvk@astro.utoronto.ca.

Abstract

We report the detection of probable optical counterparts for two Intermediate Mass Binary Pulsar (IMBP) systems, PSR J1528−3146 and PSR J1757−5322. Recent radio pulsar surveys have uncovered a handful of these systems with putative massive white dwarf companions, thought to have an evolutionary history different from that of the more numerous class of Low Mass Binary Pulsars (LMBPs) with He white dwarf companions. The study of IMBP companions via optical observations offers us several new diagnostics: the evolution of main sequence stars near the white-dwarf-neutron star boundary, the physics of white

†Part of a manuscript in preparation for publication in *The Astrophysical Journal*

dwarfs close to the Chandrasekhar limit, and insights into the recycling process by which old pulsars are spun up to high rotation frequencies. We were unsuccessful in our attempt to detect optical counterparts of PSR J1141−6545, PSR J1157−5112, PSR J1435−6100, and PSR J1454−5846.

6.1 Introduction

The majority of recycled pulsars are in low mass binary pulsar (LMBP) systems, consisting of a neutron star and a low-mass white dwarf. The LMBPs are widely considered to be descendents of the Low-Mass X-ray Binaries (LMXBs). The progenitors are thus a massive star primary (which gives rise to the neutron star) and a low mass ($\lesssim 1\,M_{\odot}$) secondary. In contrast, double neutron star binaries, exemplified by PSR B1913+16, descend from binaries in which both the primary and secondary are massive stars, each forming a neutron star.

Over the past few years, astronomers have come to appreciate the existence of another class of binary pulsars, the so-called intermediate mass binary pulsars (IMBPs) with massive C-O or O-Ne-Mg white dwarf companions. Though only 16 candidate IMBP systems are known, they may exist in numbers similar to, or perhaps even greater than, neutron star binary systems (Portegies Zwart & Yungelson, 1999).

As suggested by their name, IMPBs are thought to descend from binary star systems with a massive primary and a secondary which is intermediate in mass. The primary explodes and forms a neutron star, and at a later time, the secondary evolves into a massive white dwarf, recycling the pulsar in the process. PSR B2303+46 (Stokes et al., 1985) and PSR J1141−6545 (Kaspi et al., 2000) are two relatively young, slowly rotating binary pulsars with companion masses similar to the IMBP systems; however, in such systems it is thought that neither the primary nor the secondary is massive enough to form a neutron star. Instead, as the primary evolves it transfers matter to the secondary, thereby making the secondary massive enough to evolve into a neutron star. The final outcome is a highly eccentric system containing a massive white dwarf and a neutron star. The detection of the white dwarf companion via optical observations can help clarify this interesting evolutionary path (van Kerkwijk & Kulkarni, 1999).

Apart from these tests of binary evolution, these systems may offer us new insights into the physics of how neutron stars are spun up by accretion. It is clear that the mass transfer of the recycling process results in a decreased magnetic field, as well as an increased rotation rate for the neutron star. The spin period at the end of the spin-up phase, P_0, is a critical input to pulsar recycling models. A comparison of the white dwarf age from cooling models with the pulsar spin-down age (which assumes that P_0 is much smaller than the current spin period) can, in principle, allow the determination of P_0 (Camilo et al., 1994).

6.2 Observations

We have obtained optical observations of fields containing six IMBP systems discovered in recent radio pulsar surveys with the Parkes radio telescope (Fig. 1; Tab. 1; Camilo et al., 2001; Edwards & Bailes, 2001A; Kaspi et al., 2000; Jacoby et al., in preparation, see Chapter 3). We observed PSR J1141−6545, PSR J1157−5112, PSR J1435−6100, PSR J1528−3146, PSR J1454−5846, and PSR J1757−5322 in R band on the nights of 6 – 8 August 2002 with the the Magellan Instant Camera (MagIC) on the 6.5-m Baade telescope at Magellan Observatory. Seeing was generally good, but some targets were observed at high airmass, giving a broader point spread function. Conditions were photometric on 8 and 10 August, but there were clouds present on 9 August. Each of our six targets was observed for two 10-minute exposures on one of the photometric nights except for PSR J1528−3146. These data were reduced following standard practices (bias subtraction, flat fielding with dome flats), photometrically calibrated with observations of the Stetson standard star L112-805, and astrometrically calibrated using the USNO B-1.0 catalog. The astrometric uncertainty in all observations presented here is dominated by the tie between the USNO-B1.0 system and the International Celestial Reference Frame ($\sim 0\farcs2$ in each coordinate).

On the night of 4 June 2003, we observed PSR J1528−3146 once again with MagIC. Conditions were not photometric, but better than on our previous attempt. We obtained 2 exposures of 5 minutes each in R and 2 exposures of 10 minutes each in B, reduced in the standard manner as before. A rough photometric calibration was obtained using stars from the USNO B-1.0 catalog, which also provided the astrometric calibration.

Table 6.2 gives the relevant parameters of the best imaging observations in each band for each target. For each image, a model point spread function was constructed based on several stars in the field using the DAOPHOT package in IRAF. Limiting magnitudes were determined by placing a number of artificial stars of a given magnitude in the field and

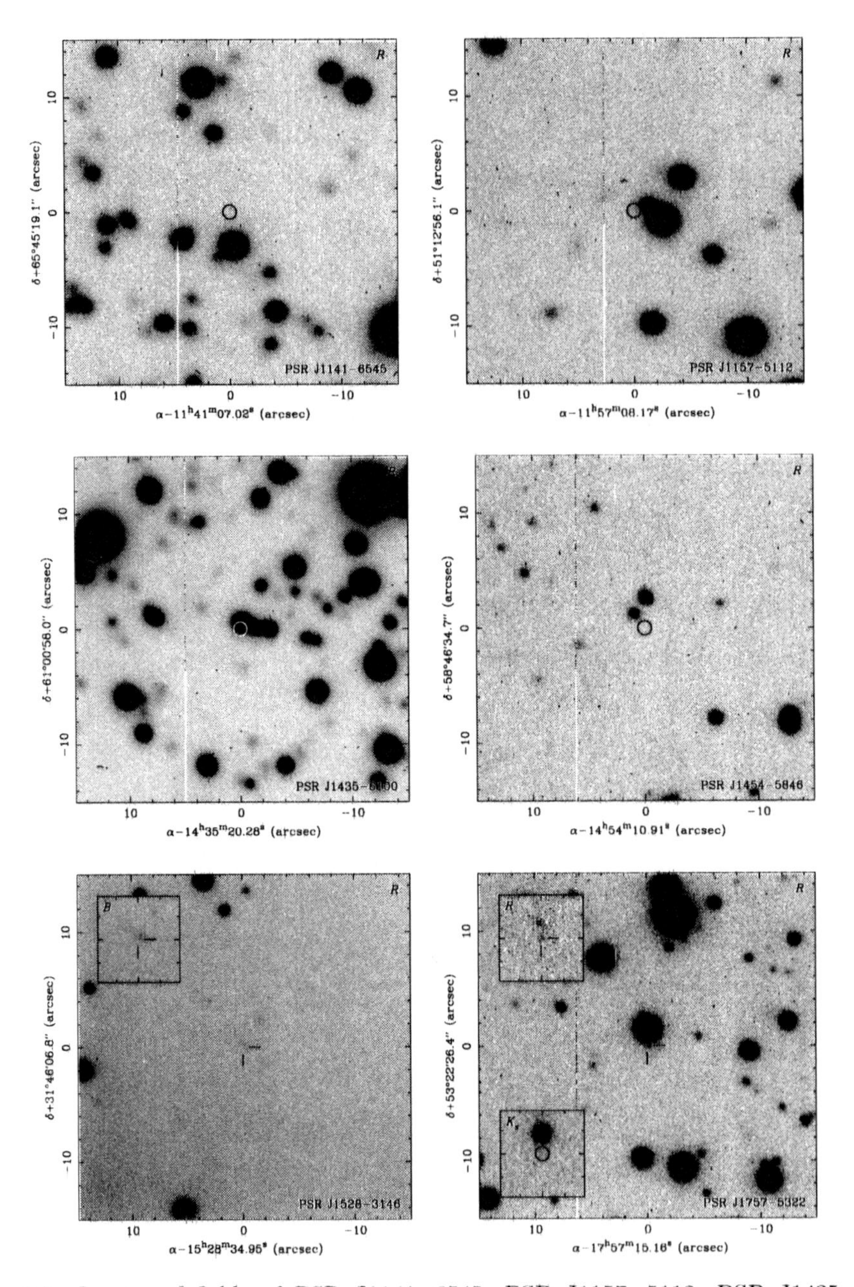

Figure 6.1: Images of fields of PSR J1141−6545, PSR J1157−5112, PSR J1435−6100, PSR J1528−3146, PSR J1454−5846, and PSR J1757−5322. Circles indicate $3\,\sigma$ uncertainty in pulsar position; tick marks show pulsar position where a plausible counterpart was detected. Large images are in R band. For PSR J1528−3146, inset shows B band image. For PSR J1757−5322, inset at upper left shows R band image after subtraction of bright star near pulsar position, and inset at lower left shows K_s image. For PSR J1141−6545, the timing position from Bailes et al. (2003) was used.

Table 6.1. Parameters of six target massive white dwarf binary systems

Pulsar	P (ms)	B (10^9 G)	τ_c (Gyr)	P_b (d)	e	$m_{c\,\min}$ ($M_\odot$)	d^a (kpc)	Reference
J1141−6545	393.9	1300	0.0014	0.20	1.8×10^{-1}	0.97	2.5	1
J1157−5112	43.6	2.5	4.8	3.51	4.0×10^{-4}	1.18	1.3	2
J1435−6100	9.3	0.5	6	1.35	1×10^{-5}	0.90	2.2	3
J1454−5846	45.2	6	0.9	12.42	1.9×10^{-3}	0.87	2.2	3
J1528−3146	60.8	3.9	3.9	3.18	2.1×10^{-4}	0.94	0.80	4
J1757−5322	8.9	0.49	5.3	0.45	$(4 \pm 4) \times 10^{-6}$	0.55	0.96	2

[a]Distance estimated from dispersion measure using model of Cordes & Lazio (2002)

References. — (1) Kaspi et al., 2000; (2) Edwards & Bailes, 2001b; (3) Camilo et al., 2001; (4) Jacoby et al., in prep.

Table 6.2. Observations of massive white dwarf binary systems

Pulsar	Filter	seeing (arcsec)	Detection Limit (magnitudes)	Potential Counterpart[a] (magnitudes)
J1141−6545	R	1.1	23.4	–
J1157−5112	R	1.2	23.7	–
J1435−6100	R	1.0	23.1	–
J1454−5847	R	0.8	24.9	–
J1528−3146	R	0.7	24.4	24.2(4)
	B	0.7	25.9	24.5(2)
J1757−5322	R	0.6	24.8	24.6(2)
	K_s	0.5	20.8	–

[a]Figures in parenthesis are uncertainties in the last digit quoted.

measuring their magnitudes with aperture photometry. This process was repeated to find the input artificial star magnitude that resulted in a standard deviation of ~ 0.3 in the measured magnitude, corresponding to a 3σ detection.

The second attempt at imaging the PSR J1528−3146 field revealed a faint object in the R band image at $\alpha_{J2000} = 15^h28^m34\overset{s}{.}955$, $\delta_{J2000} = -31°46'06\overset{''}{.}73$, and in the B band image at $\alpha_{J2000} = 15^h28^m34\overset{s}{.}945$, $\delta_{J2000} = -31°46'06\overset{''}{.}71$, consistent with the pulsar timing position. This potential counterpart is faint; we estimate $R \sim 24.2$ and $B \sim 24.5$, but this photometry is somewhat uncertain due to calibration with the USNO B-1.0 photographic magnitudes. This object is blue relative to most other stars in the field.

Our observation of PSR J1757−5322 showed a possible detection of an object at the radio pulsar's timing position, but it was difficult to see in the glare of a brighter star. Subtraction of the brighter star from the image using the DAOPHOT SUBSTAR task reveals a faint object with $R \sim 24.6$ at $\alpha_{J2000} = 17^h57^m15\overset{s}{.}174$, $\delta_{J2000} = -53^\circ22'26\overset{''}{.}17$, consistent with the pulsar timing position. We subsequently obtained a near-IR image of the field with PANIC on the 6.5 m Clay telescope at Magellan Observatory on 18 April 2003, observing for a total of 72 minutes in K_s band. We subtracted dark frames, then produced a sky frame for subtraction by taking a sliding box-car window of 4 exposures on either side of a reference exposure. We then added the exposures together, identified all the stars, and produced masks for the stars that were used to improve the sky frames in a second round of sky subtraction. Astrometry was again provided by the USNO B-1.0 catalog, and photometric calibration by comparison with several 2MASS stars in the field. There is no object present at the pulsar's position to the detection limit of the image, $K_s = 20.75$. This magnitude corresponds to a main sequence spectral type of $\sim$M4 or earlier, and is thus consistent with a white dwarf.

Several of these fields are rather crowded; this was especially problematic in the case of PSR J1435−6100, whose position overlaps with three blended objects in our image. On the night of 6 June 2003, we obtained a spectrum of the bright object near the pulsar position with the LDSS2 on the Clay telescope, and determined that it is a reddened F-type main sequence star and thus not associated with the pulsar. We used the DAOPHOT ALLSTAR task to subtract stars near the positions of PSR J1157−5112 and PSR J1435−6100, eliminating the possibility of fainter counterparts hidden by the nearby brighter objects in these cases.

6.3 Discussion and Conclusions

We detected optical counterparts for two out of the six IMBP systems we studied, PSR J1528−3146 and PSR J1757−5322. From Table 6.1, one sees that these are the two nearest targets. Thus, it is quite possible that deeper observations would reveal the counterparts in the remaining binaries as well.

In Figure 6.2, we show cooling curves for hydrogen atmosphere white dwarfs with masses from $0.5\,M_\odot$ to $1.2\,M_\odot$, along with the observationally-inferred absolute R magnitudes of massive white dwarf pulsar companions versus the spin-down ages of their pulsars. The absolute magnitudes have large uncertainties which are difficult to quantify because the only constraint on the pulsar distances is based on dispersion measure and a model of the galactic electron distribution (Cordes & Lazio, 2002); however, this exercise is still instructive. We note that in all cases where optical observations failed to detect an IMBP counterpart, the predicted magnitude is fainter than the observation's detection threshold.

As previously mentioned, it is thought that the companion stars in the PSR J1141−6545 and PSR B2303+46 systems must have been fully evolved by the time the pulsars formed. Therefore, in these systems, the pulsar age does not constrain the white dwarf age and the failure to detect the PSR J1141−6545 companion is not troubling. The detected optical counterpart of PSR B2303+46 (van Kerkwijk & Kulkarni, 1999) is significantly fainter than predicted by the cooling model based on the pulsar's spin-down age. In addition to the expectation that the white dwarf is older than the pulsar, this object has the largest z-distance from the galactic plane in this sample; it is above much of the ionized gas in the galactic disk, so the dispersion measure-based distance estimate could be significantly smaller than the true distance.

In all of the other systems, the neutron star formed first and the pulsar's spin-down age should, in principle, correspond to the time since the end of the companion's evolution. The other five detected objects are all brighter than predicted by the cooling curves if they are as old as their pulsars' characteristic ages. Although there is a large uncertainty associated with the absolute magnitude of each object, as a group, they suggest that the standard spin-down model for pulsars may in fact significantly overestimate the pulsar age in these cases, possibly because P_0 was not much smaller than the current spin period.

BAJ and SRK thank NSF and NASA for supporting their research. MHvK acknowledges support by the National Sciences and Engineering Research Council of Canada. DLK thanks the Fannie & John Hertz Foundation for its support.

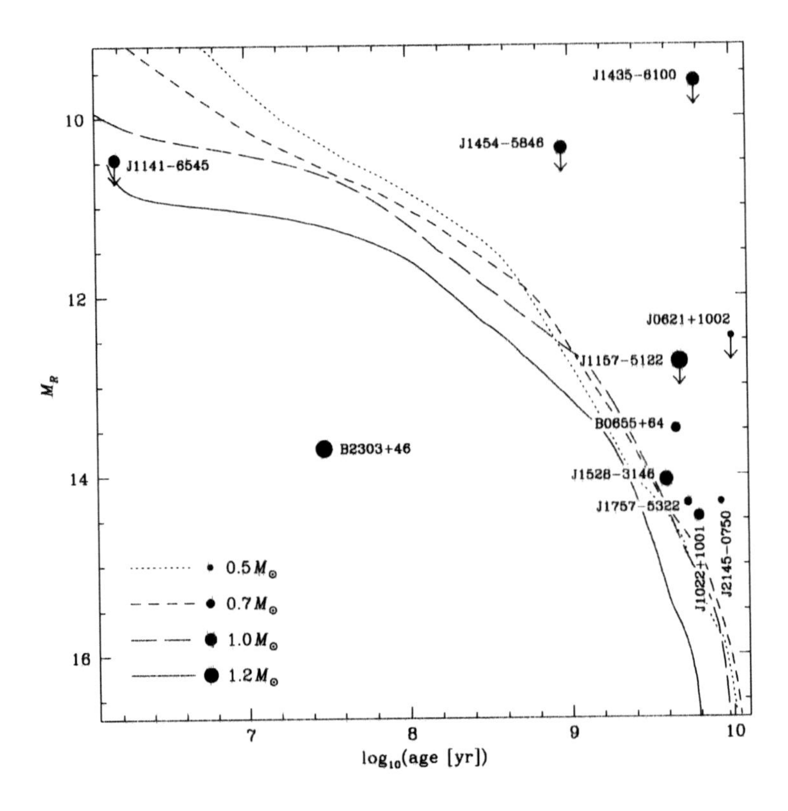

Figure 6.2: White dwarf cooling curves and observations of massive white dwarf pulsar companions. The curves show the absolute R magnitude versus age for massive white dwarfs with hydrogen atmospheres. Points show the observationally-derived M_R versus pulsar characteristic age for massive white dwarf pulsar companions with arrows indicating upper limits from non-detections. The diameter of each point is proportional to the most likely mass of the white dwarf, assuming a 1.35 $M_\odot$ pulsar and 60° orbital inclination with these exceptions: the PSR J1141−6545 companion mass measures (0.99±0.02) $M_\odot$ (Bailes et al., 2003); the most likely masses of the PSR B2303+46 and PSR J1157−5112 companions are greater than the Chandrasekhar mass, so we have assigned them diameters corresponding to 1.4 $M_\odot$. The curves are based on the luminosity – age relation for 0.5, 0.7, and 1.0 $M_\odot$ white dwarfs with 10^{-4} hydrogen fractions, and a 1.2 $M_\odot$ white dwarf with a hydrogen fraction of 10^{-6}, all with zero metallicity (Benvenuto & Althaus, 1999). To these cooling relations, we applied bolometric corrections and colors as a function of temperature for a log(g) = 8 white dwarf with hydrogen atmosphere (Bergeron et al., 1995). Apparant R magnitudes were converted to absolute R magnitudes using the dispersion measure-distance model of Cordes & Lazio (2002), with an extinction correction from Neckel & Klare (1980) for objects within 5° of the galactic plane, and from Schlegel et al. (1998) for higher latitude pulsars. In the cases of PSR J2145−0750 and PSR J0655+64, R was calculated based on the measured V and inferred temperature of Lundgren et al. (1996), using colors from Bergeron et al. (1995). Photometry for PSR J0621+1002 is from Kulkarni (1986); PSR B2303+46 from van Kerkwijk & Kulkarni (1999); and PSR J0621+1002 from van Kerkwijk et al. (2004).

Chapter 7

Measurement of Orbital Decay in the Double Neutron Star Binary PSR B2127+11C†

B. A. JACOBY[a], P. B. CAMERON[a], F. A. JENET[b], S. B. ANDERSON[a], R. N. MURTY[c], AND S. R. KULKARNI[a]

[a]Department of Astronomy, California Institute of Technology, MS 105-24, Pasadena, CA 91125; baj@astro.caltech.edu, pbc@astro.caltech.edu, sba@astro.caltech.edu, srk@astro.caltech.edu.

[b]California Institute of Technology, Jet Propulsion Laboratory, 4800 Oak Grove Drive, Pasadena, CA 91109; merlyn@alum.mit.edu.

[c]Cornell University, Ithaca, NY 14853; rnm5@cornell.edu

Abstract

We report the direct measurement of orbital period decay in the double neutron star pulsar system PSR B2127+11C in the globular cluster M15 at the rate of $(-3.95 \pm 0.13) \times 10^{-12}$, consistent with the prediction of general relativity at the $\sim 3\%$ level. We find the pulsar mass to be $m_p = (1.3584 \pm 0.0097) M_{\odot}$ and the companion mass $m_c = (1.3544 \pm 0.0097) M_{\odot}$. We also report long-term pulse timing results for the pulsars PSR B2127+11A and PSR B2127+11B, including confirmation of the cluster proper motion.

---

†Part of a manuscript in preparation for publication in *The Astrophysical Journal*

7.1 Introduction

Pulsars in binary systems with neutron star companions provide the best available laboratories for testing theories of gravity. To date, two such systems have been used for such tests: PSR B1913+16 (Taylor & Weisberg, 1982, 1989), and PSR B1534+12 (Stairs et al., 1998; Stairs et al., 2002). Both are consistent with Einstein's general relativity (GR).

The globular cluster M15 (NGC 7078) contains 8 known radio pulsars, the brightest of which are PSR B2127+11A (hereafter M15A), a solitary pulsar with a 110.6 ms spin period; PSR B2127+11B (M15B), a solitary 56.1 ms pulsar; and PSR B2127+11C (M15C), a 30.5 ms pulsar in a relativistic 8-hour orbit with another neutron star (Wolszczan et al., 1989; Anderson, 1992). The Keplerian orbital parameters of M15C are nearly identical to those of PSR B1913+16, though the former did not follow the standard high mass binary evolution (Anderson et al., 1990). With our data set spanning 12 years, M15C now provides a similar test of GR.

7.2 Observations and Analysis

We observed M15 with the 305 m Arecibo radio telescope from April 1989 to February 2001, with a gap in observations between February 1994 and December 1998 roughly corresponding to a major upgrade of the telescope. All observations used the 430 MHz line feed, with 10 MHz of bandwidth centered on 430 MHz.

Observations up to January 1994 were made with the Arecibo 3-level autocorrelation spectrometer (XCOR), which provided 128 lags in each of two circular polarizations and 506.625 μs time resolution. The autocorrelation functions were transformed to provide 128 frequency channels across the band, and these data were dedispersed at the appropriate dispersion measure (DM) (Anderson, 1992) and folded synchronously with the pulse period for each pulsar. Observations were broken into sub-integrations of 10 minutes for M15A and M15C, and 20 minutes for the fainter M15B.

Beginning in January 1999, we used the Caltech Baseband Recorder (CBR) for data acquisition. This backend sampled the telescope signal in quadrature with 2-bit resolution and wrote the raw voltage data to tape for off-line analysis with the Hewlett-Packard Exemplar machine at the Caltech Center for Advanced Computing Research (CACR). After unpacking the data and correcting for quantization effects (Jenet & Anderson, 1998), we formed a virtual 32-channel filterbank in the topocentric reference frame with each channel coherently dedispersed at the dispersion measure of M15C (Hankins & Rickett, 1975; Jenet et al., 1997). The coherent filterbank data were then dedispersed and folded for each pulsar as for the XCOR data.

The folded pulse profiles were cross-correlated against a high signal-to-noise standard profile appropriate to the pulsar and backend (Fig. 7.1) to obtain an average pulse time of arrival (TOA) for each sub-integration, corrected to UTC(NIST). The standard pulsar timing package TEMPO[1], along with the Jet Propulsion Laboratory's DE405 ephemeris, was used for all timing analysis. TOA uncertainties estimated from the cross-correlation process were multiplied by a constant determined for each pulsar-instrument pair in order to obtain reduced $\chi^2 \simeq 1$. An arbitrary offset was allowed between the XCOR and CBR data sets to account for differences in instrumental delays and standard profiles. The timing models resulting from our analysis are presented in Table 7.1, and post-fit TOA residuals relative to these models are shown in Figure 7.2.

7.3 Discussion

7.3.1 Post-Keplerian Observables for M15C

In addition to the five usual Keplerian orbital parameters, in the case of M15C we have measured three post-Keplerian (PK) parameters, advance of periastron ($\dot{\omega}$), time dilation and gravitational redshift (γ), and orbital period derivative ($\dot{P}_b$). The dependence of these PK parameters on the Keplerian parameters and component masses depend on the theory of gravity; in GR, these relations are (see Taylor & Weisberg, 1982; Damour & Deruelle, 1986; and Damour & Taylor, 1992):

$$\dot{\omega} = 3G^{2/3}c^{-2}\left(\frac{P_b}{2\pi}\right)^{-5/3}(1-e^2)^{-1}M^{2/3}, \quad (7.1)$$

$$\gamma = G^{2/3}c^{-2}e\left(\frac{P_b}{2\pi}\right)^{1/3}m_c(m_p+2m_c)M^{-4/3}, \quad (7.2)$$

$$\dot{P}_b = -\frac{192\pi}{5}G^{5/3}c^{-5}\left(\frac{P_b}{2\pi}\right)^{-5/3}(1-e^2)^{-7/2}\left(1+\frac{73}{24}e^2+\frac{37}{96}e^4\right)m_p m_c M^{-1/3}, \quad (7.3)$$

[1] http://pulsar.princeton.edu/tempo

Table 7.1. Pulsar Parameters for B2127+11A–C

Parameter[a]	Pulsar		
	B2127+11A	B2127+11B	B2127+11C
Right ascension, α_{J2000}	$21^h29^m58^s.2472(3)$	$21^h29^m58^s.632(1)$	$21^h30^m01^s.2042(1)$
Declination, δ_{J2000}	$+12^\circ10'01''.264(8)$	$+12^\circ10'00''.31(3)$	$+12^\circ10'38''.209(4)$
Proper motion in α, μ_α (mas yr^{-1})	-0.26(76)	1.7(33)	-1.3(5)
Proper motion in δ, μ_δ (mas yr^{-1})	-4.4(15)	-1.9(59)	-3.4(10)
Pulse period, P (ms)	110.66470446904(5)	56.13303552473(9)	30.52929614864(1)
Reference epoch (MJD)	50000.0	50000.0	50000.0
Period derivative, $\dot{P}$ (10^{-15})	-0.0210281(2)	0.0095406(6)	0.00498789(2)
Period second derivative, $\ddot{P}$ (10^{-30} s^{-1})	32(5)	...	-2.7(13)
Dispersion measure, DM (pc cm^{-3})	67.31	67.69	67.12
Binary model	...	...	DD
Orbital period, P_b (d)	...	...	0.33528204828(5)
Projected semimajor axis, $a \sin i$ (lt-s)	...	...	2.51845(6)
Orbital eccentricity, e	...	...	0.681395(2)
Longitude of periastron, ω (deg)	...	...	345.3069(5)
Time of periastron, T_0	...	...	50000.0643452(3)
Advance of periastron, $\dot{\omega}$ (deg yr^{-1})	...	...	4.4644(1)
Time dilation & gravitational redshift, γ (s)	...	...	0.00478(4)
Orbital period derivative, $(\dot{P}_b)^{\rm obs}$ (10^{-12})	...	...	-3.96(5)
Weighted RMS timing residual (μs)	58.9	103.5	26.0
Derived Parameters			
Orbital period derivative, $(\dot{P}_b)^{\rm int}$ (10^{-12})	...	...	-3.95(13)[b]
Pulsar mass, m_p ($M_\odot$)	...	...	1.3584(97)
Companion mass, m_c ($M_\odot$)	...	...	1.3544(97)
Total mass, $M = m_p + m_c$ ($M_\odot$)	...	...	2.71279(13)
Galactic longitude, l (deg)	65.01	65.01	65.03
Galactic latitude, b (deg)	-27.31	-27.31	-27.32
Transverse velocity, $v_\perp$ (km s^{-1})[c]	210(70)	120(230)	170(40)
Intrinsic period derivative, $\dot{P}_{\rm int}$ (10^{-15})[b]	...	...	0.00501(5)
Surface magnetic field, $B_{\rm surf}$ ($\times10^{10}$G)[b]	...	...	1.237(6)
Characteristic age, τ_c (Gyr)[b]	...	...	0.097(1)

[a]Figures in parenthesis are uncertainties in the last digit quoted. Uncertainties are calculated from twice the formal error produced by TEMPO.

[b]Corrected for kinematic effects as described in §7.3.2

[c]Based on the distance measurement of McNamara et al. (2004)

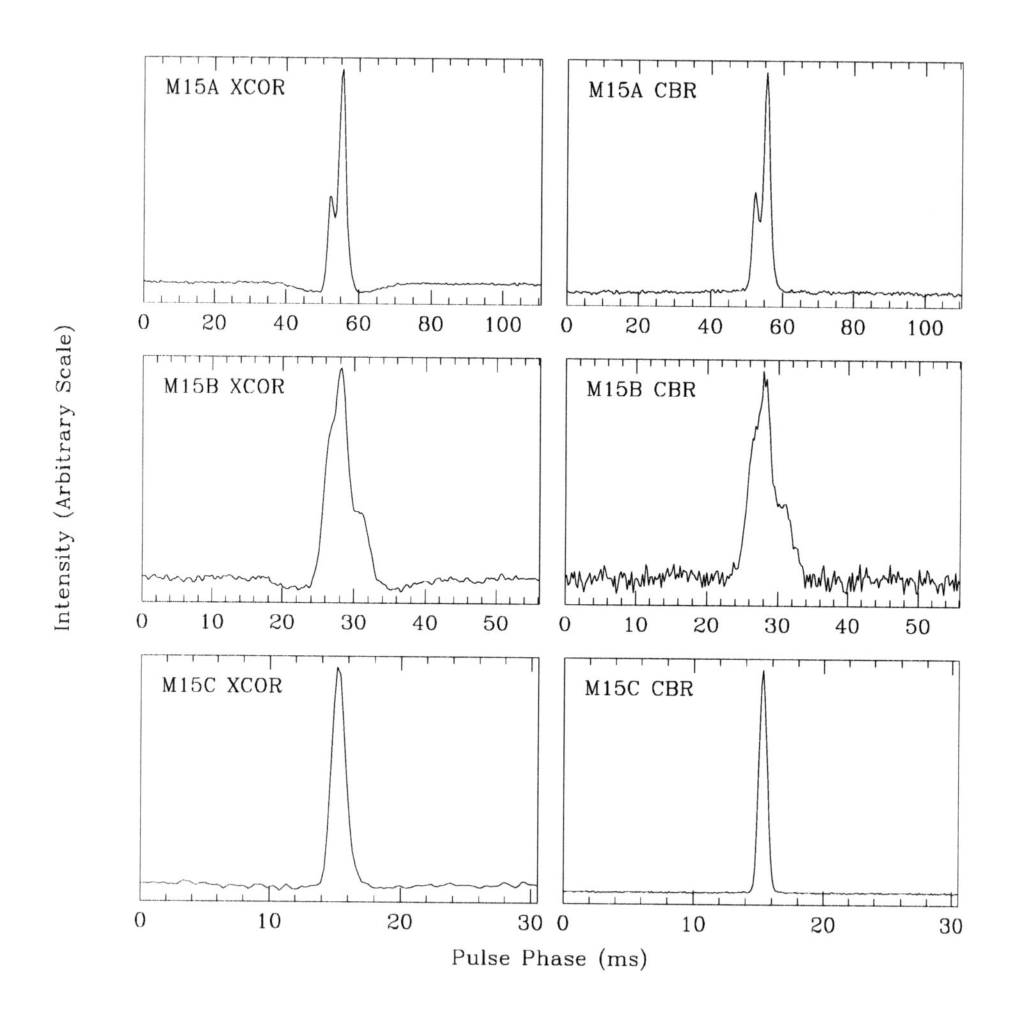

Figure 7.1: Average pulse profiles for three pulsars in M15. Average pulse profiles of M15 A (top row), M15B (middle row), and M15C (bottom row) as observed with XCOR (left column) and CBR (right column).

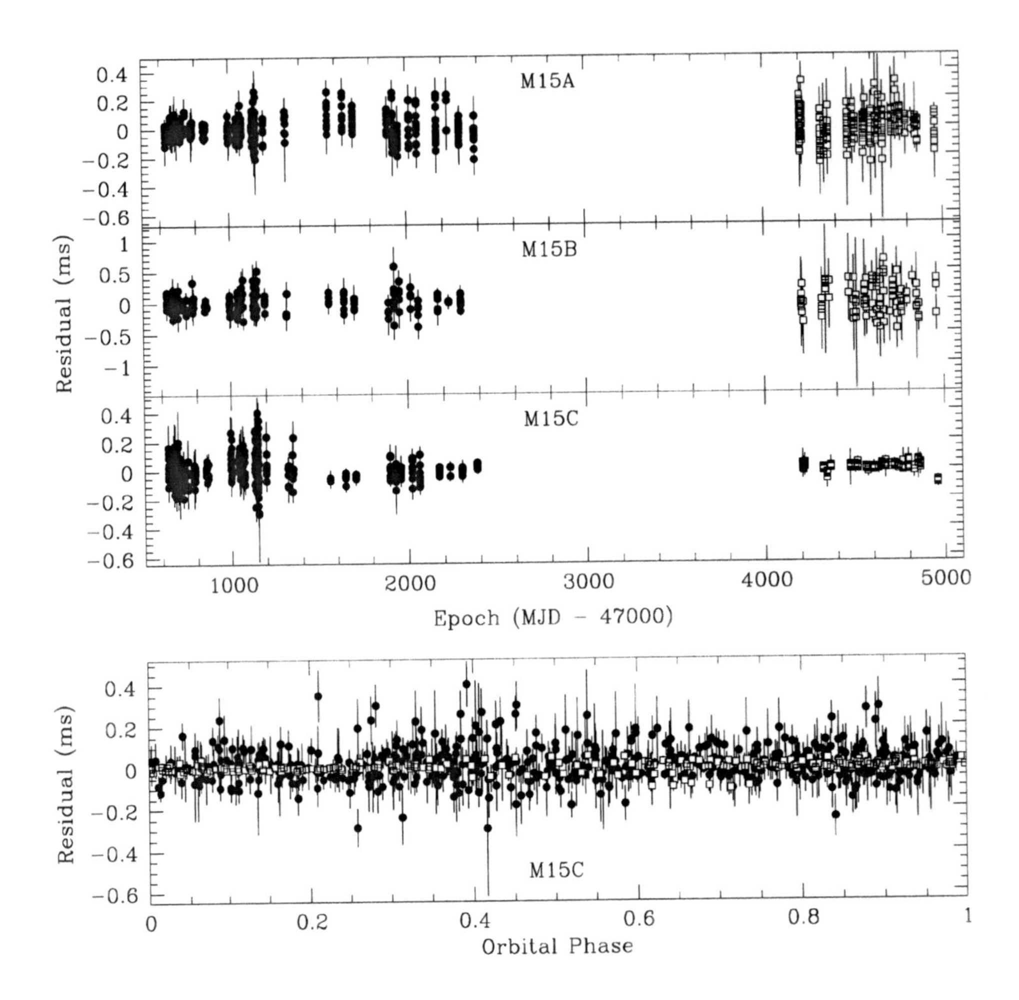

Figure 7.2: Timing residuals of M15A, M15B, and M15C. Residuals for each pulsar versus observation epoch are shown in the top three panels, with M15C residuals versus orbital phase in the bottom panel. Filled circles represent XCOR observations, with open squares indicating CBR data.

where G is the gravitational constant and c is the speed of light. The measurement of any two PK observables determines the component masses under the assumption that GR is the correct description of gravity; measuring the third parameter overdetermines the system and allows a consistency test of GR.

7.3.2 Kinematic Effects on Pulse Timing

The rate of change of orbital period that we observe in the M15C system, $(\dot{P}_b)^{\rm obs}$, is corrupted by kinematic effects that must be removed to determine the intrinsic rate, $(\dot{P}_b)^{\rm int}$. Following the discussion of Phinney (1992, 1993) regarding the parallel case of kinematic contributions to $\dot{P}$, we have

$$\left(\frac{\dot{P}_b}{P_b}\right)^{\rm kin} = -\frac{v_0^2}{cR_0}\left(\cos b\ \cos l + \frac{\delta - \cos b\ \cos l}{1+\delta^2 - 2\delta\cos b\ \cos l}\right) + \frac{\mu^2 d}{c} - \frac{a_l}{c}, \tag{7.4}$$

where $v_0 = 220 \pm 20\,{\rm km\,s^{-1}}$ is the Sun's galactic rotation velocity, $R_0 = 7.7 \pm 0.7\,{\rm kpc}$ is the Sun's galactocentric distance, $\delta \equiv d/R_0$, μ is the proper motion, $d = (9.98{\pm}0.047)$ kpc is the distance to the pulsar (McNamara et al., 2004), and a_l is pulsar's line-of-sight acceleration within the cluster. The first term in equation (7.4) is due to the pulsar's galactic orbital motion, the second to the secular acceleration resulting from the pulsar's transverse velocity (Shklovskii, 1970), and the third to the cluster's gravitational field.

Acceleration within the cluster may well dominate the kinematic contribution to $\dot{P}_b$, but a_l is an odd function of the distance from the plane of the sky containing the cluster center to the pulsar, and since we do not know if M15C is in the nearer or further half of the cluster, we must use its expectation value, $\bar{a}_l = 0$. Phinney (1993) calculates a maximum value of $|a_l|_{\rm max}/c = 6 \times 10^{-18}\,{\rm s^{-1}}$, too small for the observed $\dot{P}$ to provide a useful constraint. However, the unknown a_l still dominates the uncertainty of $(\dot{P}_b)^{\rm kin}$; we take the median value of $0.71\,|a_l|_{\rm max}$ as the uncertainty in a_l (Phinney, 1992). Evaluating equation (7.4), the total kinematic contribution is

$$\left(\dot{P}_b\right)^{\rm kin} = (-0.0095 \pm 0.12) \times 10^{-12}\,, \tag{7.5}$$

and subtracting this contamination from $(\dot{P}_b)^{\rm obs}$ yields the intrinsic value

$$\left(\dot{P}_b\right)^{\rm int} = (-3.95 \pm 0.13) \times 10^{-12}\,. \tag{7.6}$$

7.3.3 Component Masses of M15C and a Test of General Relativity

Solving equations (7.1) and (7.2) given the measured values of $\dot{\omega}$, γ, P_b, and e gives $m_p = (1.3584 \pm 0.0097)M_\odot$, $m_c = (1.3544 \pm 0.0097)M_\odot$, and $M \equiv m_p + m_c = (2.71279 \pm 0.00013)M_\odot$ in the framework of GR (Fig. 7.3). This result is consistent with, and more precise than, previous mass measurements for the neutron stars in the M15C system (Prince et al., 1991; Anderson, 1992; Deich & Kulkarni, 1996). We note that these masses are consistent with the masses of double neutron star binaries observed in the field (Thorsett & Chakrabarty, 1999; Stairs, 2004). M15 is a metal-poor cluster with a mean metallicity [Fe/H] = -2.3 (Sneden et al., 1991), suggesting that the mass of neutron stars is not a strong function of the metallicity of their progenitors.

Our determination of a third PK parameter gives a test of GR; $(\dot{P}_b)^{\rm int}$ is 1.003 ± 0.033 times the predicted value. While M15C provides an impressive test of GR, it is less stringent than the 1% $\dot{\omega}$-γ-$\dot{P}_b$ test provided by PSR B1913+16 (Taylor & Weisberg, 1989) and the 0.5% $\dot{\omega}$-γ-s test provided by PSR B1534+12 (Stairs et al., 2002), where $s \equiv \sin i$ is the shape parameter determined through measurement of Shapiro delay. We note that the uncertainty in the intrinsic orbital period decay is due almost entirely to the kinematic contribution, so further observations will not significantly improve our determination of $(\dot{P}_b)^{\rm int}$ or the quality of the test of GR it allows.

7.3.4 Proper Motion of M15

The proper motions resulting from our timing analysis give transverse velocities for M15A and M15C several times greater than the cluster escape velocity. The measured proper motions for these two pulsars and M15B are shown in Figure 7.4, along with four published proper motion measurements for M15 based on optical astrometry. The pulsar proper motions are all consistent with each other and with the cluster measurement of Cudworth & Hanson (1993).

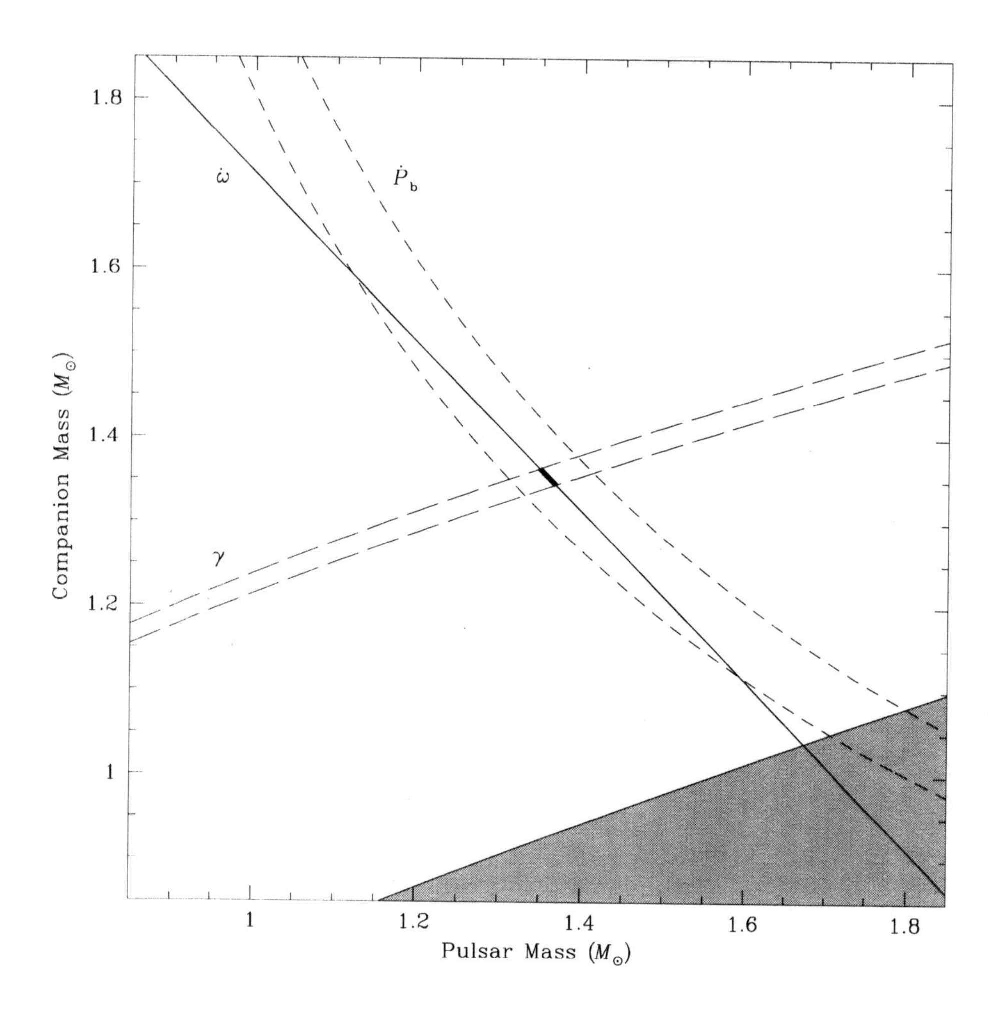

Figure 7.3: M15C mass-mass diagram. The constraints on the pulsar and companion masses in GR from the measured values of γ (long dashed lines), $\dot{\omega}$ (solid lines), and intrinsic $\dot{P}_b$ (short dashed lines) are shown. The allowed region in mass-mass space at the intersection of these constraints is denoted by a heavy line segment near the center of the plot. The shaded region is excluded by Kepler's laws. The intersection of the constraints from the three post-Keplerian observables indicates that the behavior of this system is consistent with GR.

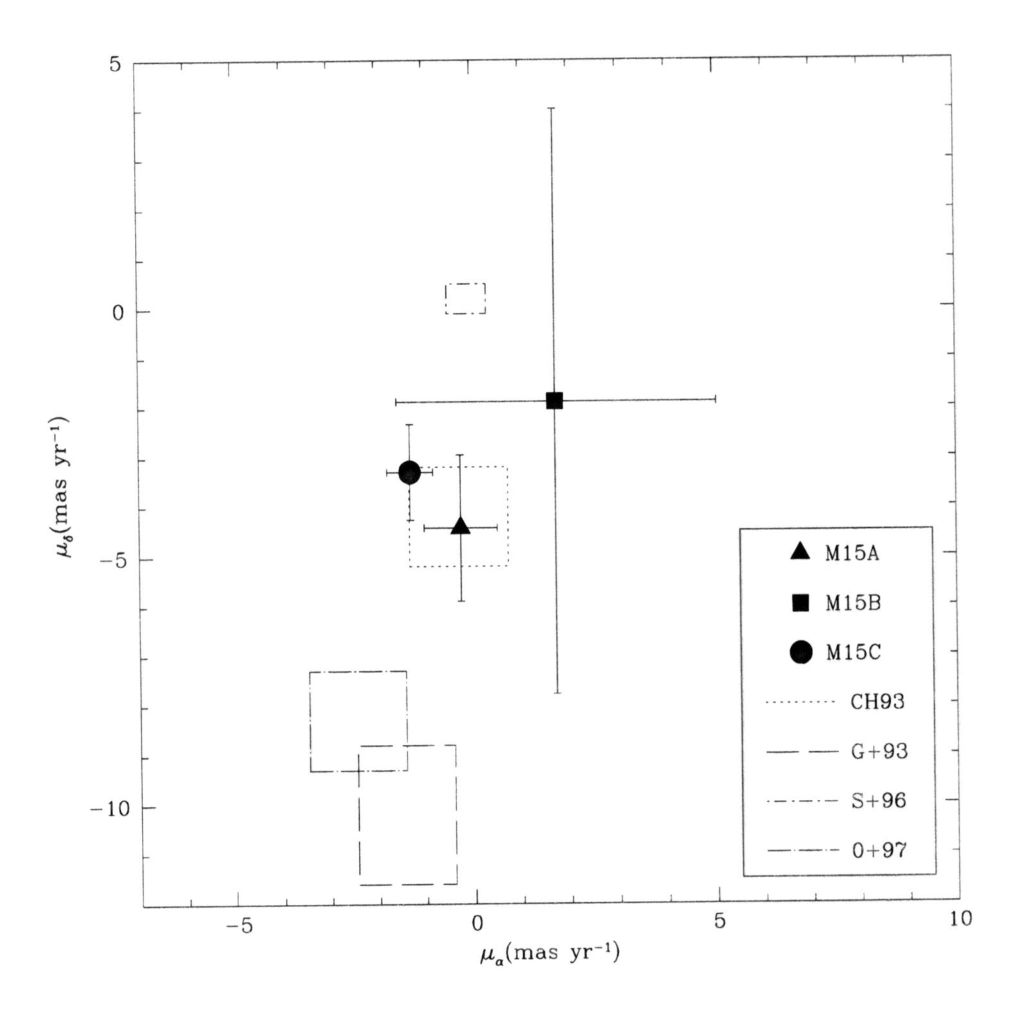

Figure 7.4: Proper motion of M15. Measured proper motions of M15A, M15B, and M15C in right ascension and declination, with rectangular regions indicating the published cluster proper motion measurements of Cudworth & Hanson (1993) (CH93), Geffert et al. (1993) (G+93), Scholz et al. (1996) (S+96), and Odenkirchen et al. (1997) (O+97).

7.3.5 Intrinsic Spin Period Derivatives

If we assume that GR provides the correct description of gravity, we can use $(\dot{P}_b)^{\rm obs}$ to determine the total kinematic correction to $\dot{P}_b$ and hence, to $\dot{P}$ for M15C. We find

$$\left(\frac{\dot{P}_b}{P_b}\right)^{\rm kin}_{\rm GR} = \left(\frac{\dot{P}}{P}\right)^{\rm kin}_{\rm GR} = (-8 \pm 17) \times 10^{-19}\,{\rm s}^{-1}\,, \tag{7.7}$$

which corresponds to $a_l/c = (4 \pm 17) \times 10^{-19}\,{\rm s}^{-1}$. We now apply this correction to the observed value of $\dot{P}$ and find the intrinsic value assuming GR, $(\dot{P})^{\rm int}_{\rm GR} = (0.00501 \pm 0.00005) \times 10^{-15}$. This intrinsic spindown rate allows us to improve upon the previous estimate of the pulsar's characteristic age and magnetic field strength (Anderson, 1992); we find $\tau_c = (0.097 \pm 0.001)$ Gyr and $B_{\rm surf} = (1.237 \pm 0.006) \times 10^{10}$ G.

Our timing models for M15A and M15C include $\ddot{P}$ (Tab. 7.1) which is unlikely to be intrinsic to the pulsars. For M15A in the cluster core, Phinney (1993) estimates the kinematic contribution to be $|\dot{a}_l/c| \equiv \left|\ddot{P}/P\right| \leq 10^{-26}\,{\rm s}^{-1}$ (80% confidence). This is significantly larger than the observed $\left|\ddot{P}/P\right| \sim 3 \times 10^{-28}\,{\rm s}^{-1}$, so the observed $\ddot{P}$ is consistent with the expected jerk resulting from the cluster's mean field and nearby stars. For M15C, far from the cluster core, we measure $\left|\ddot{P}/P\right| \sim 10^{-28}\,{\rm s}^{-1}$. We note that our measurement of $\ddot{P}$ in M15C is not of high significance ($\sim 2\sigma$), and may be an artifact of the systematic trends apparent in our timing data (Fig. 7.2).

The Arecibo Observatory, a facility of the National Astronomy and Ionosphere Center, is operated by Cornell University under cooperative agreement with the National Science Foundation. We thank W. Deich for providing the pulsar data analysis package, PSRPACK. BAJ and SRK thank NSF and NASA for supporting this research.

Appendix A

Pulsars in M62

(Jacoby et al., 2002, IAU Circular No. 7783)

B. A. Jacoby and A. M. Chandler, California Institute of Technology (Caltech); D. C. Backer, University of California, Berkeley; and S. B. Anderson and S. R. Kulkarni, Caltech, report the discovery of three new binary millisecond pulsars in the globular cluster M62 (NGC 6266): "The discovery was made with the recently commissioned 100-m Green Bank Telescope (GBT). M62 was observed at a frequency of 1.4 GHz for 4 hr on 2001 Aug. 16 with the GBT and the Berkeley-Caltech Pulsar Machine (BCPM), a flexible 2×96-channel digital filterbank. Confirmation observations were carried out during the week of Dec. 4. We searched for pulsars at the dispersion measure of the three previously known pulsars in the cluster (D'Amico et al., 2001, http://xxx.lanl.gov/abs/astro-ph/?0105122). PSR J1701-3006D has a spin period of 3.418 ms, an orbital period of 1.118 days, and a minimum companion mass of $0.12\,M_{\odot}$. PSR J1701-3006E has a 3.234-ms spin period and is orbited by a companion of at least $0.03\,M_{\odot}$ every 0.16 day. PSR J1701-3006F has a 2.295-ms spin period, an orbital period of 0.2 day, and a minimum companion mass of $0.02\,M_{\odot}$. M62 is now the third globular cluster containing six or more known radio pulsars."

Appendix B

A Wide-Bandwidth Coherent Dedispersion Backend for High-Precision Pulsar Timing

Most radio pulsars are faint radio sources with steep power law spectra, requiring observations at relatively low frequency with large fractional bandwidths to attain good sensitivity. However, the interstellar medium is dispersive, imparting a time delay of pulse arrival relative to infinite frequency,

$$\Delta t = \frac{e^2}{2\,\pi\, m_e\, c\,\nu^2}\, DM, \tag{B.1}$$

where m_e is the mass of the electron, e is the charge of the electron, ν is the observing frequency, and the dispersion measure $DM \equiv \int_0^d n_e(l)\, dl$ is equal to the integrated free electron column density along the line of sight to the pulsar. This dispersive delay must be removed from the pulsar signal to obtain optimum time resolution, and in many cases, to detect the pulsar at all. The total delay across a typical observing band is often comparable to, or even many times longer than, the period of an MSP (Fig. B.1).

Traditionally, dedispersion has been accomplished by dividing the observing band into a number of frequency channels, detecting the power in each channel as a function of time, and shifting the channels relative to each other before summing to remove the dispersive delay. This channelization is often accomplished with an analog or digital filterbank, an autocorrelation spectrometer, or a hardware-based fast Fourier transform engine (see Jacoby & Anderson, 2001 for a review). However, with these post-detection dedispersion techniques, there is no way to eliminate the differential delay within a frequency channel, which means that dedispersion is always imperfect and the time resolution of a given channel is

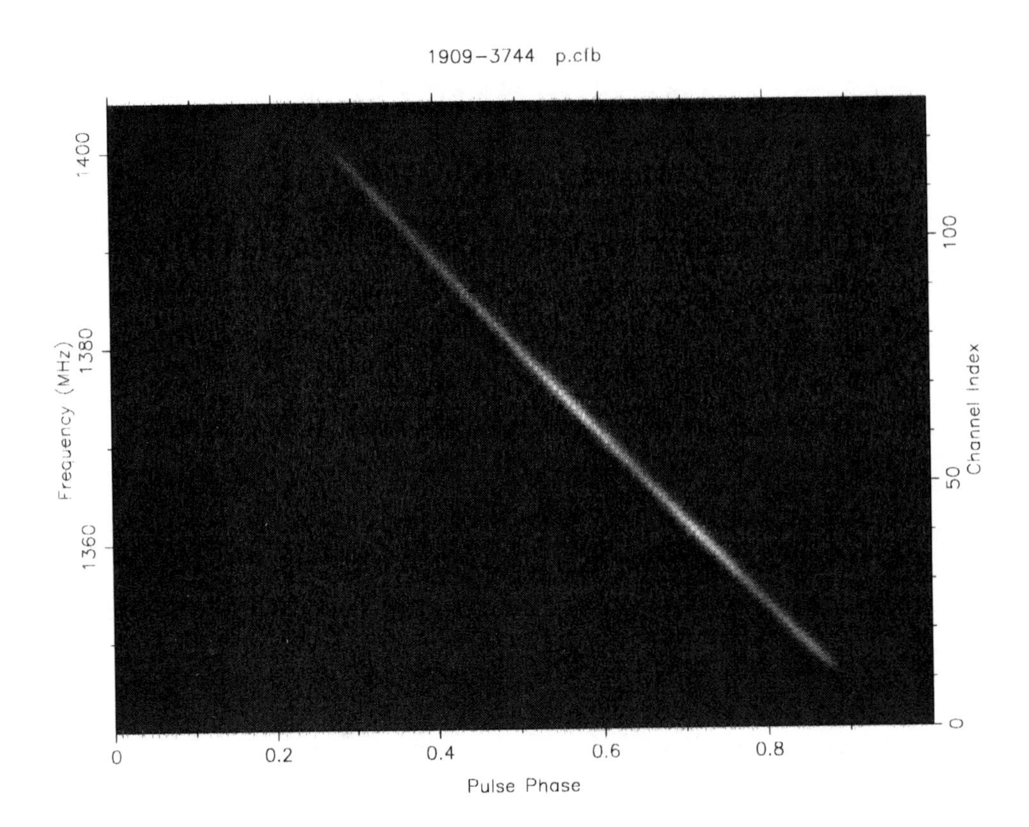

Figure B.1: Dispersed pulse profile. The propagation speed of the pulsar signal is slower at lower frequencies, causing a frequency-dependent delay (Equation B.1) that must be removed to attain optimal sensitivity. This observation of PSR J1909−3744 was made with the CGSR2 coherent dedispersion backend at the 100-m Green Bank Telescope, splitting the sampled 64 MHz-wide band into 128 coherently-dedispersed channels. Rolloff at the edges of the band is due to anti-aliasing filters; brightness fluctuations within the band are due to scintillation.

limited to its inverse bandwidth. There is a tradeoff between time resolution and frequency resolution, making this an imperfect solution.

Another solution is pre-detection, or coherent dedispersion. In this technique, the digitized voltage signal from the telescope receiver is convolved with the inverse of the dispersive transfer function (Hankins & Rickett, 1975), most often through multiplication in the Fourier domain. The application of this technique has until recently been limited to relatively narrow bandwidths by the huge data volumes generated by sampling the telescope signal at the Nyquist rate, and the daunting computing task of processing the data (e.g. Stairs et al., 2000).

We have constructed the first wide-bandwidth near-real-time coherent dedispersion backend in the form of a second-generation fast flexible digitizer (FFD2) mated to a dedicated cluster of computers (Fig. B.2). The digitizer was designed at built at the California Institute of Technology by J. K. Yamasaki with assistance from J. Maciejewski, S. B. Anderson, B. A. Jacoby, and S. R. Kulkarni. This system provides four 64 MHz-wide bands at 2-bit resolution (128 MHz total bandwidth in each of two polarizations), or two 64 MHz bands at 4-bit resolution. The computing architecture and software were designed and implemented by M. Bailes, S. Ord, A. Hotan, and W. van Straten at Swinburne University of Technology.

The FFD2 (Fig. B.3) has four analog intermediate frequency (IF) input channels, each with two variable gain amplifiers to allow for optimum pre-digitization level setting and dynamic range. The board was designed with DC offset adjustment capability in mind, but this has proved unnecessary and has not been implemented. The analog signals are sampled at 16-bit resolution at a frequency of 128 MHz, driven by a dedicated off-board clock tied to the observatory frequency standard, with the start of data acquisition synchronized with a 1 pulse per second (PPS) signal also synchronized to the observatory standard. A single field-programmable gate array (FPGA) chip handles all control and data handling operations, packing samples from the four input channels (or two channels in 4-bit resolution mode) into 16-bit words, output through low-voltage differential signaling (LVDS) interfaces to a pair of off-the-shelf EDT PCI-CD60 data acquisition cards. A header with incrementing counter is periodically inserted into the data stream to ensure data synchronization. The FFD2 is controlled via serial connection. The sampler board is housed in a VME-format crate, but interfaces with a custom backplane.

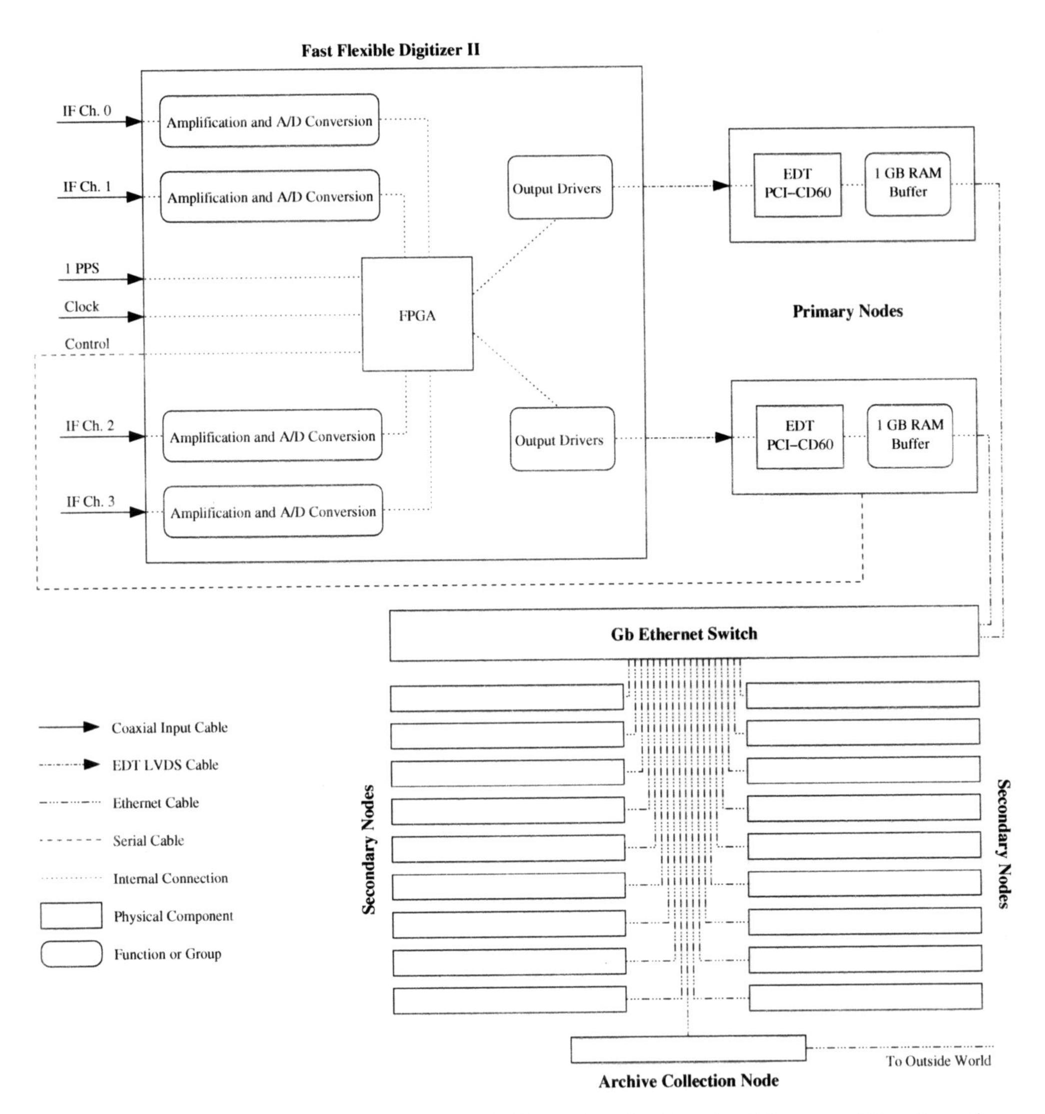

Figure B.2: Block diagram of coherent dedispersion backend. After amplification, four band-limited IF signals from telescope are digitized synchronously with 128 MHz sample clock, with the start of data taking signaled by 1 PPS tick. Digital data are packed by FPGA and sent to EDT PCI-CD60 data acquision cards in primary node computers via LVDS interface. Data are distributed to seconday nodes for processing via Gb ethernet, and final folded profile archives are collected by a computer which also serves as the network gateway.

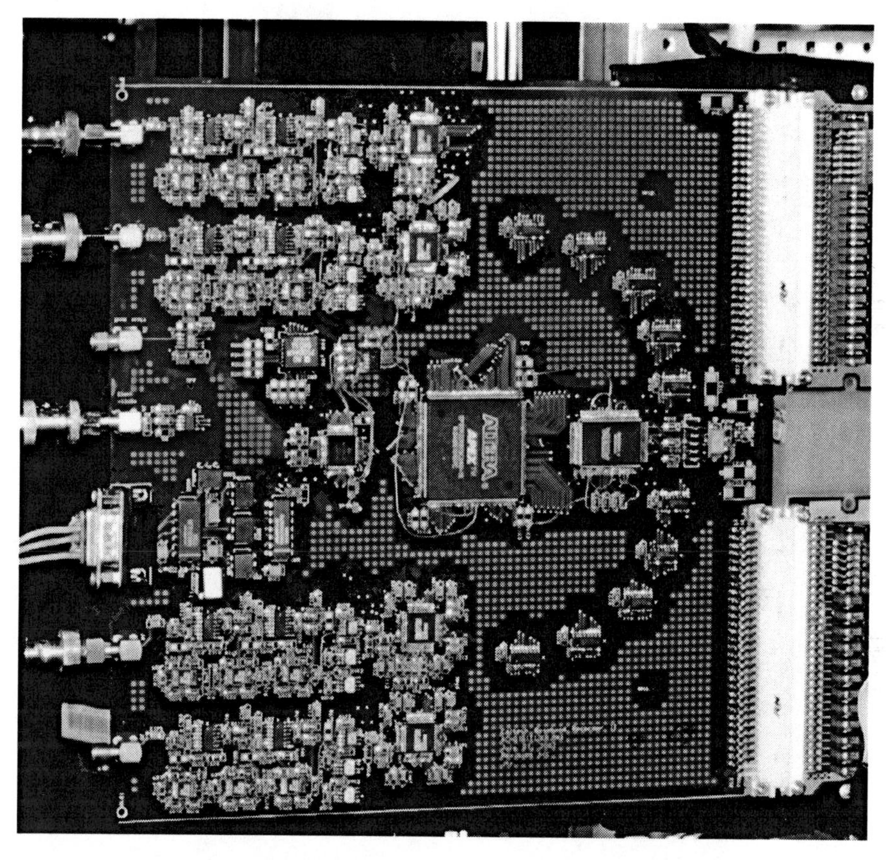

Figure B.3: Fast Flexible Digitizer II. Connectors on the left side of the board from top to bottom are IF inputs 0 and 1, 1 PPS start signal, 128 MHz sampler clock, serial control connection, and IF inputs 2 and 3. The two large connectors on the right handle LVDS data output as well as power supply.

Figure B.4: Caltech – Green Bank – Swinburne Recorder II. The crate containing the FFD2, IF filters and amplifiers, 128 MHz sampler clock, and power supplies is at the bottom of the rack, below the dedicated computing cluster and Gb ethernet switch.

The input frequency range of each channel is limited to 64 MHz to 128 MHz with external bandpass filters. This half-Nyquist sampling aliases the input band to the DC – 64 MHz range. Real sampling is employed to avoid possible systematic errors arising in the mixing process required for complex sampling.

Each of the two EDT data acquisition cards installed in dual-processor Intel Xeon computers receive data at an externally-clocked rate of 64 MB s^{-1}, slightly in excess of their nominal maximum of 60 MB s^{-1}. The data from each of these two channels are then distributed in 1 GB (roughly 17 s) segments via standard Gb ethernet to a cluster of computers for processing or storage. Generally, the data are unpacked and corrected for quantization effects (Jenet & Anderson, 1998) and immediately reduced using the PSRDISP coherent dedispersion software package (van Straten, 2002), but they can also be stored on disk for other processing strategies such as pulsar search or single pulse analysis. This flexible computing strategy is described more fully by Bailes (2003).

The first of these systems, the Caltech – Parkes – Swinburne Recorder II (CPSR2), was installed at the 64-m Parkes radio telescope in August, 2002 and began regular science observations in December, 2002. In January 2004, the Caltech – Green Bank – Swinburne Recorder II (CGSR2) was installed at the 100-m Green Bank Telescope and began regular observations in July, 2004. These two instruments are nearly identical, differing mainly in the specifications of their respective computer clusters, as well as IF amplification and anti-aliasing filters as needed for each telescope.

The author's role in the development of these systems included contributions to FFD2 design review and troubleshooting (including writing test and evaluation software), installation, and commissioning observations. The author also had primary responsibility for specifying and procuring the computing equipment, sampler clock, and anti-aliasing filters for the CGSR2 system.

We thank the staff of the Parkes Observatory and the NRAO – Green Bank for their support in the installation and operation of these instruments. The Parkes telescope is part of the Australia Telescope which is funded by the Commonwealth of Australia for operation as a National Facility managed by CSIRO. The National Radio Astronomy Observatory is a facility of the National Science Foundation operated under cooperative agreement by Associated Universities, Inc.

Bibliography

Anderson, S. B. 1992, Ph.D. thesis, California Institute of Technology

Anderson, S. B., Gorham, P. W., Prince, T. A., Kulkarni, S. R., & Wolszczan, A. 1990, Nature, 346, 42

Baade, W. & Zwicky, F. 1934, Publications of the National Academy of Sciences, 20, 254

Backer, D. C., Kulkarni, S. R., Heiles, C., Davis, M. M., & Goss, W. M. 1982, Nature, 300, 615

Bailes, M. 2003, in ASP Conference Series, Vol. 302, Radio Pulsars, ed. M. Bailes, D. J. Nice, & S. E. Thorsett (San Francisco: Astronomical Society of the Pacific), 57

Bailes, M., Ord, S. M., Knight, H. S., & Hotan, A. W. 2003, ApJ, 595, L49

Bassa, C. G., van Kerkwijk, M. H., & Kulkarni, S. R. 2003, A&A, 403, 1067

Bell, J. F. 1998, Advances in Space Research, 21, 137

Bell, J. F. & Bailes, M. 1996, ApJ, 456, L33

Benvenuto, O. G. & Althaus, L. G. 1999, MNRAS, 303, 30

Bergeron, P., Wesemael, F., & Beauchamp, A. 1995, PASP, 107, 1047

Bildsten, L. 2003, in ASP Conference Series, Vol. 302, Radio Pulsars, ed. M. Bailes, D. J. Nice, & S. E. Thorsett (San Francisco: Astronomical Society of the Pacific), 289

Burderi, L., Possenti, A., D'Antona, F., Di Salvo, T., Burgay, M., Stella, L., Menna, M. T., Iaria, R., Compana, S., & D'Amico, N. 2001, ApJ, 560, L71

Burgay, M., D'Amico, N., Possenti, A., Manchester, R. N., Lyne, A. G., Joshi, B. C., McLaughlin, M. A., Kramer, M., Sarkissian, J. M., Camilo, F., Kalogera, V., Kim, C., & Lorimer, D. R. 2003, Nature, 426, 531

Camilo, F., Lyne, A. G., Manchester, R. N., Bell, J. F., Stairs, I. H., D'Amico, N., Kaspi, V. M., Possenti, I., Crawford, F., & McKay, N. P. F. 2001, ApJ, 548, L187

Camilo, F., Nice, D. J., & Taylor, J. H. 1993, ApJ, 412, L37

—. 1996, ApJ, 461, 812

Camilo, F., Thorsett, S. E., & Kulkarni, S. R. 1994, ApJ, 421, L15

Chakrabarty, D., Morgan, E. H., Muno, M. P., Galloway, D. K., Wijnands, R., van der Klis, M., & Markwardt, C. B. 2003, Nature, 424, 42

Cordes, J. M. & Chernoff, D. F. 1997, ApJ, 482, 971

Cordes, J. M. & Lazio, T. J. W. 2002, astro-ph/0207156

Cudworth, K. M. & Hanson, R. B. 1993, AJ, 105, 168

D'Amico, N., Possenti, A., Manchester, R., Sarkissian, J., Lyne, A., & Camilo, F. 2001, astro-ph/0105122

Damour, T. & Deruelle, N. 1985, Ann. Inst. H. Poincaré (Physique Théorique), 43, 107

—. 1986, Ann. Inst. H. Poincaré (Physique Théorique), 44, 263

Damour, T. & Taylor, J. H. 1992, Phys. Rev. D, 45, 1840

Deich, W. T. S. & Kulkarni, S. R. 1996, in Compact Stars in Binaries: IAU Symposium 165, ed. J. van Paradijs, E. P. J. van del Heuvel, & E. Kuulkers (Dordrecht: Kluwer), 279

Driebe, T., Schoenberner, D., Bloecker, T., & Herwig, F. 1998, A&A, 339, 123

Edwards, R. T. & Bailes, M. 2001a, ApJ, 547, L37

—. 2001b, ApJ, 553, 801

Edwards, R. T., Bailes, M., van Straten, W., & Britton, M. C. 2001, MNRAS, 326, 358

Foster, R. S., Cadwell, B. J., Wolszczan, A., & Anderson, S. B. 1995, ApJ, 454, 826

Freire, P. C., Camilo, F., Kramer, M., Lorimer, D. R., Lyne, A. G., Manchester, R. N., & D'Amico, N. 2003, MNRAS, 340, 1359

Geffert, M., Colin, J., Le Campion, J.-F., & Odenkirchen, M. 1993, AJ, 106, 168

Hankins, T. H. & Rickett, B. J. 1975, in Methods in Computational Physics Volume 14 — Radio Astronomy (New York: Academic Press), 55

Hewish, A., Bell, S. J., Pilkington, J. D. H., Scott, P. F., & Collins, R. A. 1968, Nature, 217, 709

Hobbs, G., Faulkner, A., Stairs, I. H., Camilo, F., Manchester, R. N., Lyne, A. G., Kramer, M., D'Amico, N., Kaspi, V. M., Possenti, A., McLaughlin, M. A., Lorimer, D. R., Burgay, M., Joshi, B. C., & Crawford, F. 2004, MNRAS, 352, 1439

Hulse, R. A. & Taylor, J. H. 1975, ApJ, 195, L51

Jacoby, B. A. 2003, in ASP Conference Series, Vol. 302, Radio Pulsars, ed. M. Bailes, D. J. Nice, & S. E. Thorsett (San Francisco: Astronomical Society of the Pacific), 133

Jacoby, B. A. & Anderson, S. B. 2001, in Encyclopedia of Astronomy and Astrophysics, ed. P. Murdin (Bristol: Institute of Physics), 795

Jacoby, B. A., Bailes, M., van Kerkwijk, M. H., Ord, S., Hotan, A., Kulkarni, S. R., & Anderson, S. B. 2003, ApJ, 599, L99

Jacoby, B. A., Chandler, A. M., Backer, D. C., Anderson, S. B., & Kulkarni, S. R. 2002, IAU Circ. No. 7783

Jaffe, A. H. & Backer, D. C. 2003, ApJ, 583, 616

Jenet, F. A. & Anderson, S. B. 1998, PASP, 110, 1467

Jenet, F. A., Cook, W. R., Prince, T. A., & Unwin, S. C. 1997, PASP, 109, 707

Kaspi, V. M., Lyne, A. G., Manchester, R. N., Crawford, F., Camilo, F., Bell, J. F., D'Amico, N., Stairs, I. H., McKay, N. P. F., Morris, D. J., & Possenti, A. 2000, ApJ, 543, 321

Kaspi, V. M., Taylor, J. H., & Ryba, M. 1994, ApJ, 428, 713

Kulkarni, S. R. 1986, ApJ, 306, L85

Lange, C., Camilo, F., Wex, N., Kramer, M., Backer, D., Lyne, A., & Doroshenko, O. 2001, MNRAS, 326, 274

Lattimer, J. M. & Prakash, M. 2001, ApJ, 550, 426

Lomen, A. N., Backer, D. C., & Nice, E. M. S. D. J. 2003, in ASP Conference Series, Vol. 302, Radio Pulsars, ed. M. Bailes, D. J. Nice, & S. E. Thorsett (San Francisco: Astronomical Society of the Pacific), 81

Lommen, A. N., Zepka, A., Backer, D. C., McLaughlin, M., Cordes, J. M., Arzoumanian, Z., & Xilouris, K. 2000, ApJ, 545, 1007

Lundgren, S. C., Foster, R. S., & Camilo, F. 1996, in ASP Conf. Ser. 105: IAU Colloq. 160: Pulsars: Problems and Progress, ed. S. Johnston, M. A. Walker, & M. Bailes (San Francisco: Astronomical Society of the Pacific), 497

Lyne, A. G., Burgay, M., Kramer, M., Possenti, A., Manchester, R. N., Camilo, F., McLaughlin, M. A., Lorimer, D. R., D'Amico, N., Joshi, B. C., Reynolds, J., & Freire, P. C. C. 2004, Science, 303, 1153

Lyne, A. G., Manchester, R. N., Lorimer, D. R., Bailes, M., D'Amico, N., Tauris, T. M., Johnston, S., Bell, J. F., & Nicastro, L. 1998, MNRAS, 295, 743

Manchester, R. N., Lyne, A. G., Camilo, F., Bell, J. F., Kaspi, V. M., D'Amico, N., McKay, N. P. F., Crawford, F., Stairs, I. H., Possenti, A., Morris, D. J., & Sheppard, D. C. 2001, MNRAS, 328, 17

Manchester, R. N., Lyne, A. G., D'Amico, N., Bailes, M., Johnston, S., Lorimer, D. R., Harrison, P. A., Nicastro, L., & Bell, J. F. 1996, MNRAS, 279, 1235

McLaughlin, M. A., Lorimer, D. R., Champion, D. J., Arzoumanian, Z., Backer, D. C., Cordes, J. M., Fruchter, A. S., Lommen, A. N., & Xilouris, K. M. 2004, astro-ph/0404181

McNamara, B. J., Harrison, T. E., & Baumgardt, H. 2004, ApJ, 602, 264

Monet, D. G., Levine, S. E., Canzian, B., Ables, H. D., Bird, A. R., Dahn, C. C., Guetter, H. H., Harris, H. C., Henden, A. A., Leggett, S. K., Levison, H. F., Luginbuhl, C. B., Martini, J., Monet, A. K. B., Munn, J. A., Pier, J. R., Rhodes, A. R., Riepe, B., Sell, S., Stone, R. C., Vrba, F. J., Walker, R. L., Westerhout, G., Brucato, R. J., Reid, I. N., Schoening, W., Hartley, M., Read, M. A., & Tritton, S. B. 2003, AJ, 125, 984

Neckel, T. & Klare, G. 1980, A&AS, 42, 251

Nicastro, L., Lyne, A. G., Lorimer, D. R., Harrison, P. A., Bailes, M., & Skidmore, B. D. 1995, MNRAS, 273, L68

Nice, D., Splaver, E. M., & Stairs, I. H. 2003, in ASP Conference Series, Vol. 302, Radio Pulsars, ed. M. Bailes, D. J. Nice, & S. E. Thorsett (San Francisco: Astronomical Society of the Pacific), 75

Nice, D. J., Fruchter, A. S., & Taylor, J. H. 1995, ApJ, 449, 156

Nice, D. J., Splaver, E. M., & Stairs, I. H. 2004, astro-ph/0411207

Nice, D. J., Taylor, J. H., & Fruchter, A. S. 1993, ApJ, 402, L49

Odenkirchen, M., Brosche, P., Geffert, M., & Tucholke, H. J. 1997, New Astronomy, 2, 477

Ord, S. M., van Straten, W., Hotan, A. W., & Bailes, M. 2004, MNRAS, 352, 804

Phinney, E. S. 1992, Phil. Trans. Roy. Soc. Lond. A, 341, 39

Phinney, E. S. 1993, in Structure and Dynamics of Globular Clusters, ed. S. G. Djorgovski & G. Meylan (San Francisco: Astronomical Society of the Pacific), 141

Portegies Zwart, S. F. & Yungelson, L. R. 1999, MNRAS, 309, 26

Prince, T. A., Anderson, S. B., Kulkarni, S. R., & Wolszczan, W. 1991, ApJ, 374, L41

Ransom, S. M. 2001, Ph.D. thesis, Harvard University

Ray, P. S., Thorsett, S. E., Jenet, F. A., van Kerkwijk, M. H., Kulkarni, S. R., Prince, T. A., Sandhu, J. S., & Nice, D. J. 1996, ApJ, 470, 1103

Rohrmann, R. D., Serenelli, A. M., Althaus, L. G., & Benvenuto, O. G. 2002, MNRAS, 335, 499

Schlegel, D. J., Finkbeiner, D. P., & Davis, M. 1998, ApJ, 500, 525

Scholz, R.-D., Odenkirchen, M., Hirte, S., Irwin, M. J., Borngen, F., & Ziener, R. 1996, MNRAS, 278, 251

Sheinis, A. I., Bolte, M., Epps, H. W., Kibrick, R. I., Miller, J. S., Radovan, M. V., Bigelow, B. C., & Sutin, B. M. 2002, PASP, 114, 851

Shemar, S. L. & Lyne, A. G. 1996, MNRAS, 282, 677

Shklovskii, I. S. 1970, Sov. Astron., 13, 562

Sneden, C., Kraft, R. P., Prosser, C. F., & Langer, G. E. 1991, AJ, 102, 2001

Staelin, D. H. & Reifenstein, III, E. C. 1968, Science, 162, 1481

Stairs, I. H. 2004, Science, 304, 547

Stairs, I. H., Arzoumanian, Z., Camilo, F., Lyne, A. G., Nice, D. J., Taylor, J. H., Thorsett, S. E., & Wolszczan, A. 1998, ApJ, 505, 352

Stairs, I. H., Splaver, E. M., Thorsett, S. E., Nice, D. J., & Taylor, J. H. 2000, MNRAS, 314, 459

Stairs, I. H., Thorsett, S. E., Taylor, J. H., & Wolszczan, A. 2002, ApJ, 581, 501

Staveley-Smith, L., Wilson, W. E., Bird, T. S., Disney, M. J., Ekers, R. D., Freeman, K. C., Haynes, R. F., Sinclair, M. W., Vaile, R. A., Webster, R. L., & Wright, A. E. 1996, Publications of the Astronomical Society of Australia, 13, 243

Stokes, G. H., Taylor, J., & Dewey, R. J. 1985, ApJ, 294, L21

Taylor, J. H. 1974, A&AS, 15, 367

Taylor, J. H. & Weisberg, J. M. 1982, ApJ, 253, 908

—. 1989, ApJ, 345, 434

Thorsett, S. E. & Chakrabarty, D. 1999, ApJ, 512, 288

Toscano, M., Bailes, M., Manchester, R., & Sandhu, J. 1998, ApJ, 506, 863

Toscano, M., Sandhu, J. S., Bailes, M., Manchester, R. N., Britton, M. C., Kulkarni, S. R., Anderson, S. B., & Stappers, B. W. 1999, MNRAS, 307, 925

van Kerkwijk, M. & Kulkarni, S. R. 1999, ApJ, 516, L25

van Kerkwijk, M. H., Bassa, C. G., Jacoby, B. A., & Jonker, P. G. 2004, astro-ph/0405283

van Kerkwijk, M. H., Bergeron, P., & Kulkarni, S. R. 1996, ApJ, 467, L89

van Straten, W. 2002, ApJ, 568, 436

van Straten, W., Bailes, M., Britton, M., Kulkarni, S. R., Anderson, S. B., Manchester, R. N., & Sarkissian, J. 2001, Nature, 412, 158

Weisberg, J. M. & Taylor, J. H. 2003, in ASP Conference Series, Vol. 302, Radio Pulsars, ed. M. Bailes, D. J. Nice, & S. E. Thorsett (San Francisco: Astronomical Society of the Pacific), 93

Wolszczan, A., Kulkarni, S. R., Middleditch, J., Backer, D. C., Fruchter, A. S., & Dewey, R. J. 1989, Nature, 337, 531

Zou, W. Z., Wang, N., Wang, H. X., Manchester, R. N., Wu, X. J., & Zhang, J. 2004, MNRAS, 354, 811

CPSIA information can be obtained at www.ICGtesting.com
Printed in the USA
LVOW111245200613

339270LV00004B/347/P

9 781581 123937